Food Safety Management

Food Safety Management

Edited by **Margo Field**

SYRAWOOD
PUBLISHING HOUSE

New York

Published by Syrawood Publishing House,
750 Third Avenue, 9th Floor,
New York, NY 10017, USA
www.syrawoodpublishinghouse.com

Food Safety Management
Edited by Margo Field

International Standard Book Number: 978-1-68286-003-8 (Hardback)

The publisher's policy is to use permanent paper from mills that operate a sustainable forestry policy. Furthermore, the publisher ensures that the text paper and cover boards used have met acceptable environmental accreditation standards.

Trademark Notice: Registered trademark of products or corporate names are used only for explanation and identification without intent to infringe.

Printed in the United States of America.

Contents

Preface

This book was inspired by the evolution of our times; to answer the curiosity of inquisitive minds. Many developments have occurred across the globe in the recent past which has transformed the progress in the field.

Food safety management as a discipline is concerned with the regulation of food production and storage processes in order to prevent potential health hazards and infections from contaminated food products. This book outlines the processes and applications of food safety management in detail with concepts such as different bacterial and viral pathogens, environmental contaminants, pesticides and drugs, food sampling, evaluation and analysis, etc. It contains contributions of internationally acclaimed scholars. The chapters included herein make this book an essential guide for both professionals and those who wish to pursue this discipline further.

This book was developed from a mere concept to drafts to chapters and finally compiled together as a complete text to benefit the readers across all nations. To ensure the quality of the content we instilled two significant steps in our procedure. The first was to appoint an editorial team that would verify the data and statistics provided in the book and also select the most appropriate and valuable contributions from the plentiful contributions we received from authors worldwide. The next step was to appoint an expert of the topic as the Editor-in-Chief, who would head the project and finally make the necessary amendments and modifications to make the text reader-friendly. I was then commissioned to examine all the material to present the topics in the most comprehensible and productive format.

I would like to take this opportunity to thank all the contributing authors who were supportive enough to contribute their time and knowledge to this project. I also wish to convey my regards to my family who have been extremely supportive during the entire project.

Editor

Contamination status and health risk assessment of trace elements in foodstuffs collected from the Buriganga River embankments, Dhaka, Bangladesh

Faysal Elahi Khan[1], Yeasmin Nahar Jolly[2*], GM Rabiul Islam[1], Shirin Akhter[2] and Jamiul Kabir[2]

Abstract

Background: Unsafe food consumption is a severe problem because of heavy metal contamination, which is caused by director indirect activities of industries. The present study was conducted to assess the risk of human health by Heavy metals (Cu, Co, Fe, Zn and Mn) through the intake of vegetables and fishes obtained from the area adjacent to the Hazaribag tannery campus, Dhaka, Bangladesh.

Result: The trend of mean metal concentration in Buriganga river water was Fe >Mn > Zn > Cu > Co and according to Department of Environment, Dhaka Bangladesh (DoE) (1999) the value of the above metals are within the permissible limit of irrigation water except Fe. An assessment of risk involved due to consumption of contaminated food also calculated. The trend of metals in vegetables was Fe > Mn > Zn > Cu > Co and in fishes the trend was Fe > Zn >Mn > Co > Cu. Accumulation of trace elements in vegetables was lower than maximum tolerable levels proposed by FAO/WHO food standard programme (2001) with the exception of Fe and Co respectively. In fishes metal concentration was lower than safe limit set by WHO (1989) except Mn. The Metal Pollution Index (MPI) for all the foodstuffs showed a higher value, however the calculated Health Risk Index (HRI) indicated no risk to human health upon consumption of those foodstuffs.

Conclusion: The overall study suggests that foodstuff in the area were contaminated by the assayed metals and long-term consumption can cause potential health risks to consumers.

Keywords: Daily intake; Heavy metal; Health risk; Metal pollution index

Background

Rapid development, growing populations, as well as increasing urbanization and food demand in Southeast Asia combined with the use of polluted waters for pre-urban food constitute a potential food safety risk. The knowledge of dietary intake of essential and toxic elements in Southeast Asian countries such as Bangladesh is limited. The poisoning effects of heavy metals are due to their interference with normal body biochemistry in normal metabolic process (Okunola et al. 2011). Heavy metals are classified in two main categories i.e. essential and toxic heavy metals. Essential heavy metals (Cu, Co, Zn, Fe, Ca, Mg, Se, Ni and Mn) are required in very trace quantities for the proper functioning of enzyme systems, hemoglobin

formation and vitamin synthesis in human but metabolic disturbances are encountered in case of excess of these essential metals (Hina et al. 2011). In recent time, the rapid and exponential industrial growth in Bangladesh caused increased production of waste, which witnessed in Hazaribag industrial area located towards southwestern Dhaka. In Hazaribag area, there are about 277 tannery industries, 15 dying, 3 salt industries, 1 pharmaceutical industry, 1 soap factory, 2 match factories and 2 lead-zinc industries. Buriganga River, which is the major sources of water supply for agricultural, livestock and fishing activities (Azom et al. 2012), is contaminated by the wastes discharged from these industries as it contains a bulk amount of liquid and solid wastes with substantial quantities of heavy metals: Zn, Cu, Co, Fe, Mn, Ca, Ni, Mg, Se etc. The contamination of river allows these pollutants to accumulate in common fish species by biomagnifications, which are used as local food sources (Azom et al. 2012). Long-term

* Correspondence: jolly_tipu@yahoo.com
[2]Chemistry Division, Atomic Energy Centre, Dhaka, Bangladesh
Full list of author information is available at the end of the article

use of wastewater in irrigation affect food quality thus safety (Muchuweti et al. 2006 and Sharma et al. 2007). Plants growing in nearby zone of industrial areas display increased concentration of heavy metals serving in many cases as bio-monitors of pollution loads (Mingorance et al. 2007). Vegetables cultivated in soil polluted by toxic heavy metals due to industrial activities take up heavy metals and accumulated them in their edible and non-edible parts.

Heavy metal pollution is of significant ecological or environmental concern because they are not easily biodegradable or metabolized thus precipitating far-reaching effects on the biological system such as human, animals, plants and other soil biota (Yoon 2003). Food chain contamination is the major pathway of heavy metal exposure for humans (Khan et al. 2008). In the present study area there are more than 0.2 million people and 20,000 people are directly exposed to the hazards (Asaduzzaman et al. 2002) and rest of people are affected by Dietary intake which is the main route of exposure of heavy metals for most people (Tripathi et al. 1997). Thus, information about heavy metal concentration in food products and their dietary intake is very important for assessing their risk to human health (Zhuang et al. 2009). The present study aimed to assess the contamination status of trace elements in water, vegetables and fishes of the adjacent area of the Hazaribag and embank of Buriganga River, Dhaka. The concentrations of trace elements in water, vegetables and fishes were compared with the established safe limit and the value of intake metals in human diet was calculated to estimate the risk to human health.

Methods
Study area
The study was carried out in the cluster of tannery industries in the Hazaribagh area of southwestern Dhaka. It lies within longitude $90°22'$-$90°22'48''$E and latitude $23°43'34''$ $23°43'49''$ N and located by the side of River Buriganga. Lot of industries suited in Hazaribag and most of them discharge their effluents without any prior treatment in the Buriganga River. Different kinds of vegetables such as Brinjal, Eggplants, Chili, Stem Amaranth, Radish, Spinach etc. are cultivated in the nearby land of these industries throughout the year. Beside, a variety of small fishes also cultured in the Buriganga River throughout the year. A map of the sampling site is shown in Figure 1.

Collection and preparation of samples
Water, vegetables viz., Stem Amaranth (*Amaranthus lividus*), Radish (*Raphanus sativus*), Spinach (Spinacia oleracea); fishes viz., Tatina (*Cirrhinus reba*), Spotted snakehead (*Channa punctate*), MozambiqueTalipia (*Oreochromis mossambicus*) were randomly collected in triplicate from different location of the study area during

the period of February to March 2013. The samples were tagged and carried in the laboratory.

Preparation of water sample
For Energy Dispersive X-ray Fluorescence (EDXRF) measurement of elements, the preparation of water samples involved the absorption of a certain amount of the sample on weighed amount of dry analar grade Whatman cellulose powder. A volume of 500 ml of each collected water sample filtered with Whatman 41 filter paper was taken in a clean weighed porcelain dish followed by addition of 4 gm of cellulose powder and evaporated on water bath. The sample after evaporation to dry mass was further dried under IR lamp at about 70°C for two hours to remove the trace of moisture and weighed. For homogeneous mixing, the dry mass was then transferred to a carbide mortar and ground to fine powder using a pestle. The processed sample in a plastic vial with identification mark was preserved inside a desiccator.

Preparation of plant and fish samples
The plant samples were cut into suitable pieces with a stainless steel knife, washed first with tap water, and rinsed with deionized water three times. The inedible parts of all fish samples were removed with a stainless steel knife. The remaining edible part of the samples were washed with tap water and then rinsed with deionized water three times. All plant and fish samples were then taken on porcelain dishes separately. Each dish with the particular sample was marked by an identification number and placed in an oven at around 70°C for overnight drying which was continued until a constant weight was obtained. The dried mass of each sample was then transferred to a carbide mortar and ground to fine powder using a pestle and preserved in a plastic vial with identification mark inside a desiccator.

Analysis of samples
The Panalytical Epsilon 5 Energy Dispersive X-ray Fluorescence (EDXRF) (model: Epsilon 5, made in Netherland) was used as major analytical technique for carrying out elemental analysis in the samples. For irradiation of the sample with X-ray beam 2 g of each powdered material was pressed into a pellet of 25 mm diameter with a pellet maker (CARVER, model: 3889-4NE1, U.S.A.) and loaded into the X-ray excitation chamber with the help of automatic sample changer system. The irradiation of all real samples was performed by assigning a time-based programme, controlled by a software package provided with the system. The standard materials were also irradiated under similar experimental conditions for construction of the calibration curves for quantitative elemental determination in the respective

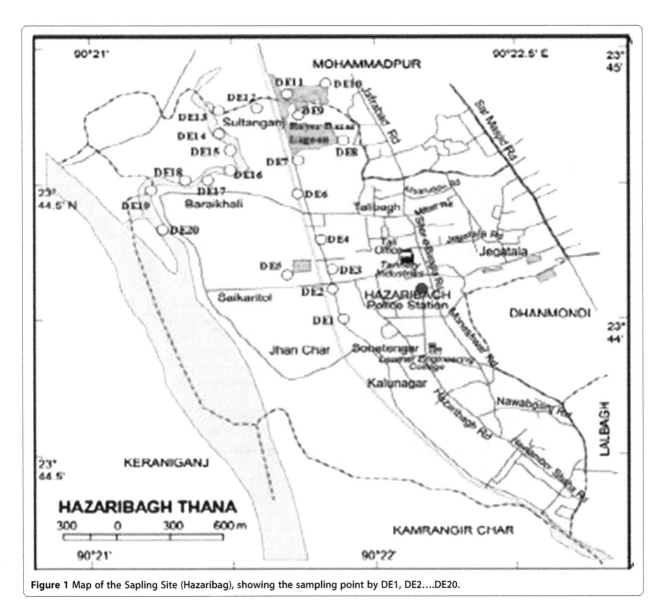

Figure 1 Map of the Sapling Site (Hazaribag), showing the sampling point by DE1, DE2....DE20.

samples. The generated X-ray spectra of the materials were stored into the computer. The X-ray intensities of the elements in sample spectrum were calculated using the system software by integration of area of the respective X-ray peak areas using peak fitting deconvolution software.

Concentration calibration and method validation

A direct comparison method based on EDXRF technique was used for elemental concentration measurement Jolly et al. (2013). As the analysis is based on direct comparison, the standards of similar matrices were used for the construction of the calibration curve in order to avoid the matrix effect. Three lab-synthesized cellulose-based multi element standards, five plant standards (Apple Leaf/NIST 1516, Spinach/NIST 1570a, Orchard Leaf/NIST 1571, Tomato Leaf/NIST 1573a, Peach Leaf/NIST

1574) and three fish standards (Tuna-1, Tuna-2, Tuna-3) were used for the construction of calibration curves for carrying out elemental analysis in river water, plant and fish samples respectively. The calibration curve for each element was constructed based on the K X-ray and L X-ray intensities calculated for the respective elements present in standard samples. The curves were constructed by plotting the sensitivities of the elements as a function of their atomic number. The validation of calibration curves constructed for elements present in the standards were checked through analysis of standard reference materials (Tuna fish for fish sample and Spinach for vegetable samples). The results obtained for elements of interest and certified values for corresponding elements are shown in the Table 1. All results in respect of certified known values were found to vary within the acceptable range of error.

Table 1 Comparison between present results and the certified values of standard reference materials (mg kg⁻¹)

Elements	Tuna Fish			Spinach		
	Results Obtained	Certified Values	Error	Results Obtained	Certified Values	Error
K	-	-	-	27729	29030	2.26
Ca	9119	9640	5.41	14483	15270	4.92
Cr	-	-	-	-	-	-
Mn	2181	2140	−1.92	69.33	75.90	0.76
Fe	39341	43200	8.93	-	-	-
Ni	8.44	8.00	−5.50	-	-	-
Cu	3405	3420	0.44	13.30	12.20	−8.98
Zn	4197	4180	−0.41	-	-	-
As	1420	1540	7.79	0.035	0.038	9.21
Se	1.05	1.00	−4.50	0.053	0.050	6.00
Pb	5443	5520	1.39	-	-	-

Statistical analysis

To assess the contamination level of heavy metals, mean, median, minimum, maximum, and standard deviation of water, fish and vegetable samples were performed using Microsoft Excel (version 2007).

Data analysis

Metal Pollution Index (MPI)

Metal pollution index (MPI) was computed to determine overall trace elements concentrations in different foodstuff analysed. This index was obtained by calculating the geometrical mean of concentrations of all the metals in different foodstuff (Ureso et al. 1997)

$$MPI(\mu g \cdot g^{-1}) = (Cf_1 \times Cf_2 \times ... \times Cf_n)^{1/n}$$

Where Cf_n = concentration of metal in n in the sample.

Health Risk Index (HRI)

The health risk index was calculated as the ratio of estimated exposure of test vegetables and fishes and oral reference dose (Cui et al. 2004). Oral reference doses were 4×10^{-2} and 0.3 mg kg⁻¹ day⁻¹ for Cu and Zn respectively (USEPA 1989); 10–60 and 0.5-5.0 mg kg⁻¹ day⁻¹ for Fe and Mn respectively (Friberg et al. 1984) and 3.01 mg kg⁻¹ day⁻¹ for Co (Food and Nutritional Board 2004). In the present study the maximum limit for Mn and Fe for oral reference dose is considered for HRI calculation. Estimated exposure is obtained by dividing daily intake of heavy metals by their safe limit. An index more than 1 is considered as not safe for human health (USEPA 2002).

Daily intake was calculated by the following equation

$$\text{Daily intake of metal (DIM)} = \frac{C_{metal} \times D_{food\ intake}}{B_{average\ weight}}$$

Where, C_{metal}, $D_{food\ intake}$, and $B_{average\ weight}$ represent the heavy metal concentrations in foodstuff ($\mu g\ g^{-1}$), daily intake of foodstuff and average body weight, respectively. According to the food consumption, survey by Alam et al. (2003) and Kennedy et al. (2001), Bangladeshi people per person per day consumes vegetable: 130 g and fish: 24 g The average body weight ($B_{average\ weight}$) was taken as 70 kg for adults according to WHO 1993.

Result and discussions

Heavy metal concentration in Buriganga river water

In contaminated water, the concentration of Fe was found highest (31.091 mg/l) and ranged from 20.683 to 39.313 mg/l which was much higher than the range 4.13 to 5.53 mg/l, reported by Ahmad and Goni (2010) and lowest concentration was found for Co (0.014 mg/l). Copper concentration ranged from 0.025 to 0.081 mg/l, which was lower than the values (0.07-6.30 mg/l) reported by Gupta et al. (2008). Maximum Zn concentration in Buriganga river water was 0.427 mg/l. Mn concentration was varied from <0.74 to 1.56 mg/l which was higher than the values (0.031-0.017 mg/l) reported by Samir et al. (2008). Co concentration was varied from 0.005 to 0.019 mg/l which was lower than the values (8.08-3.68 mg/l) reported by Muwanga and Barifaijo (2006). Zn concentration was varied from 0.167 to 0.427 mg/l, which was higher than the values (0.135-

Table 2 Heavy metal concentration (mg/l) in Buriganga river water

Metals	Safe limit[a]	Mean	Median	Max	Min	S.D
Cu	0.20	0.058	0.067	0.081	0.025	0.022
Co	0.05	0.014	0.016	0.019	0.005	0.005
Fe	5	31.091	34.330	39.313	20.683	7.885
Zn	2	0.234	0.187	0.427	0.167	0.109
Mn	0.2	1.560	1.56	1.56	<0.74	0.698

[a]safe limit of toxic heavy metals in irrigation water for agricultural purpose (Pescod 1992).

0.111 mg/l) reported, by Muiruri et al. (2013). The mean concentration (mg/l) of five heavy metals Cu, Co, Fe, Zn and Mn in water samples was 0.058, 0.014, 31.091, 0.234 and 1.560 respectively (Table 2) and according to Pescod (1992) all the metals are within the safe limit for water to be used in irrigation except Fe and Mn. In a study, Gerbrekidan et al. (2013) also found the concentration of heavy metals (Cu, Zn, Fe, Mn, Cr, Cd, Ni, Co and Pb) in the water of Gifel River near Shaba Tannery, Tigray, Northern Ethiopia, were lower than the permissible limit allowed for irrigation water. The higher standard deviation observed for heavy metals Cu, Fe, Zn and Mn in river water which may be due to the ununiform distribution of the respective metals from effluent of various industries. Sharma et al. (2006) also observed a similar trend in the wastewater of Varanasi, India. The data obtained for heavy metals in water from the present study varied more or less regularly with the findings of the other authors (Khan et al. 1998; Al-Nakshabandi et al. 1997). This variation might be ascribed to a variety of industries discharging their treated and/or untreated waste water in to the Buriganga river .hence the concentration of heavy metals in industrial effluents depends on the process of product manufacturing and raw materials used in the industries (Sharma et al. 2006).

Heavy metal concentration in vegetables

Alam et al. (2003) reported, the average per capita consumption of leafy and non-leafy vegetables is 130 g person^{-1} day^{-1} in Bangladesh, whis is considerably less than the recommended amounts of 200 g person^{-1} day^{-1} from nutritional point of view (Hasan and Ahmad 2000). The range and mean concentration of heavy metal (mg/kg dry weight) in edible parts of vegetables grown in the adjacent area of Hazaribag industrial area, Dhaka, Bangladesh are shown in Table 3. In leafy vegetable (Spinach), the Cu concentration ranged from 3.58 to 6.72 mg/kg which was higher than the value reported in Tehran, Iran (0.13 to 0.37 mg/kg) by Delbari and Kulkarni (2013) but lower than the value (15.9-17.4 mg/kg) reported by Arora et al. (2008). In stem Amaranth, mean Cu concentration was 8.12 mg/kg, which was higher than the value 4.87 mg/kg reported by Jolly et al. (2013). In radish Cu concentration ranged 3.58 to 4.32 mg/kg which was higher than the value (0.346-0.389 mg/kg) reported by Farooq et al. (2008).The higher Cu levels in the studied vegetables may be due to the elevated uptake of the heavy metals by plants grown in adjacent area of Hazaribag industrial region of Bangladesh. Concentration of Co was varied from 0.53 to 1.44 mg/kg and found in the order of Stem Amaranth > Spinach > Radish. In the present study, the variation of Co concentration in vegetables was strongly supported by the findings (1.03-1.57 mg/kg) of Naser et al. (2011). Mean Fe

Table 3 Heavy metal concentration (mg/kg dry weight) in vegetables grown in and around Hazaribag tannery industrial region, Dhaka, Bangladesh

Elements	Stem Amaranth (n = 13)	Radish (n = 10)	Spinach (n = 8)	Safe limit[a]	WAV*
Cu	8.12 (4.40-14.38)	4.02 (3.58-4.32)	5.64 (3.58-6.72)	73	5-30
Co	1.44 (0.45-3.16)	0.53 (0.36-0.65)	1.36 (0.96-1.66)	50	0.02-1
Fe	1375 (192–3489)	305 (180–381)	1186 (904–1345)	425	-
Zn	24.34 (0–44.53)	13.57 (7.43-23.21)	78.34 (44.66-98.53)	100	27-150
Mn	29.57 (0–80.15)	<0.63 (86–299)	211.56	500	30-300

n: number of samples.
[a]safe limit (joint FAO/WHO food standard programme 2001).
*WAV = World Average Value (Pendias 2000).

concentration varied from 305 to 1375 mg/kg, which was higher than the concentration (111–378 mg/kg), observed in vegetables by Arora et al. (2008). Maximum uptake of Fe was found in Stem Amaranth (1375 mg/kg), followed by Spinach (1186 mg/kg) and Radish (305 mg/kg). Ahmad and Goni (2010) also reported a higher Fe concentration in vegetables. The highest concentration of Zn was found in Spinach (78.34 mg/kg) followed by Stem Amaranth (24.34 mg/kg) and Radish (13.57 mg/kg) which agreed with the findings of Liu et al. (2005) in the vegetables from Beijing, China (32.01-69.26 mg/kg) but substantially lower (3.00-171.03 mg/kg) than the vegetables from Titagarh, West Bengla, India (Gupta et al. 2008); Harare, Zimbabwe (1038–1872 m/kg) (Tandi et al. 2004). The maximum concentration of Mn was exhibited by Spinach (211.56 mg/kg) followed by Stem Amaranth (29.57 mg/kg) and Radish (<0.63 mg/kg). Concentration of Mn in the studied vegetables was ranged from <0.63-211.56 mg/kg which is higher than the values (11.97-22.09 mg/kg) reported by Ahmad and Goni (2010), (61.86-156.24 mg/kg) reported by Jan et al. (2011). However concentration of metals in all the vegetable analysed were within in the safe limit suggested by FAO/WHO, (2001) except Fe and within the word average value (Pendias and Pendias 2000). The exhibition of spatial and temporal variations of all the heavy metal concentration might be ascribed to the variations in heavy metal sources and the quantity of heavy metals discharged through the effluents and sewage in irrigation water of the study area.

Heavy metal concentration in fishes

In a study Hossain et al. (2008) showed that according to international standard the average per capita consumption of fish is 49 g/person/day but he found 24 g/person/day in Bangladesh. The range and mean concentration of heavy metals (mg/kg dry weight) in fishes are presented in Table 4. In fresh water fish MozambiqueTalipia, concentration of Cu ranged from 2.92 to 3.24 mg/kg which was higher than the value reported in Malaysia (0.27-0.35 mg/kg) (Mokhtar et al. 2009) but lower than the value (4.03-5.57) reported by Ahmad et al. (2010). In Tatina, mean Cu concentration was 3.15 mg/kg, which was lower than the value 5.09 mg/kg reported by Ahmad et al. (2010). In Spotted snakehead mean Cu concentration was 2.84 mg/kg, which was lower than the value 5.27 mg/kg reported Ahmad et al. (2010). Concentration of Co varied from 8.17 to 9.09 mg/kg and found in the order of Spotted snakehead > Tatina > MozambiqueTalipia. Mean Fe concentration varied from 83.48 to 217.49 mg/kg which was higher than the concentration (0.3-0.35 mg/kg) observed in fishes in Aba River, Nigeria by Ubalua et al. (2007). Fe concentration (194–235 mg/kg) in Mozambique Talipia was higher than the value reported in Asa River, Ilorin, Nigeria (2.95-8.59 mg/kg) (Eletta et al. 2003). Concentration of Fe in fish samples were found in the order of MozambiqueTalipia > Tatina > Spotted snakehead. The highest concentration of Zn was found in Tatina (30.12 mg/kg) followed by Spotted snakehead (24.70 mg/kg) and MozambiqueTalipia (24.52 mg/kg). Concentration of Zn (24.52-30.122 mg/kg) in the studied fishes were higher than the value reported in Aba river, Nigeria (1.5-2.5 mg/kg) (Ubalua et al. 2007), but substantially lower (9.8-1226.40 mg/kg) than the concentrations in fresh water fishes of northern delta lakes, Egypt (Saeed and Shaker 2008). The maximum concentration of Mn was exhibited by Spotted snakehead (11.75 mg/kg) followed by MozambiqueTalipia (10.35 mg/kg) and Tatina (6.35 mg/kg). Concentration of Mn (6.35-11.75 mg/kg) in fishes were higher than the values (0.18-0.22 mg/kg) reported by Ubalua et al. (2007) but lower than the value reported by Muiruri et al. (2013) in Athi river, Nigeria (147.72-149.7 mg/kg). In the present study Cu, Zn, Fe

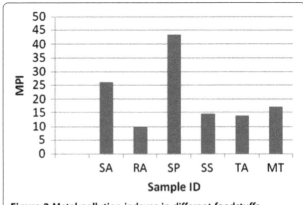

Figure 2 Metal pollution indexes in different foodstuffs analysed. SA=Stem Amaranth; RA=Radish; SP=Spinach; SS= Spotted snakehead; TA= Tatina; MT= Mozambique Talipia.

Concentration in fishes was within the safe limit of WHO except Mn. The exhibition of spatial and temporal variations of all the heavy metal concentration might be ascribed to the variations in heavy metal sources and the quantity of heavy metals discharged through the effluents and sewage in Buriganga River. Although metal concentration depends on the fish's uptake ability, accumulation ability, fish age, species etc.

Metal pollution index

Metal pollution index (MPI) is suggested to be a reliable and precise method for metal pollution monitoring of wastewater irrigation areas (Ureso et al. 1997). Among different vegetables and fishes metal pollution index (MPI) as shown in Figure 2 followed a decreasing sequence of Spinach > Stem Amaranth > Mozambique Talipia > Spotted snakehead > Tatina > Radish. Leafy vegetables are found to contain higher MPI value than fishes which are may be due to the uptake of higher amount of heavy metal available from polluted soil. This value also agrees with the findings of Singh et al. (2010) of waste water irrigated site in north east Varanasi and the sequences was as Spinach > Lady's Finger > Tomato > Brinjal > Radish. However higher MPI in Spinach, Stem

Table 4 Heavy metal concentration (mg/kg) in Fishes cultured nearby the area of tannery complex in the river Buriganga

Metals	Spotted snakehead (n = 9)	Tatina (n = 9)	MozambiqueTalipia (n = 9)	Safe limit
Cu	2.84 (2.77-2.93)	3.15 (3.02-3.28)	3.05 (2.92-3.24)	30
Co	9.09 (9.01-9.14)	9.04 (9–9.08)	8.17 (7.96-8.44)	-
Fe	83.48 (83.15-83.89)	93.53 (93.25-93.96)	217.49 (194–235)	100
Zn	24.70 (24.42-24.98)	30.12 (3.02-3.28)	24.52 (24.37-24.64)	100
Mn	11.75 (7.74-16.27)	6.35 (6.28-6.43)	10.35 (10.03-10.75)	1

n: number of samples.
Safe limit: WHO (1989).

Table 5 Health risk index (HRI) of heavy metals via intake if foodstuffs from the adjacent area of Hazaribag

Foodstuffs	HRI				
	Cu	Co	Fe	Zn	Mn
Stem Amaranth	0.38	8.88 E-3	0.04	1.13	1.09 E-1
Radish	0.19	3.27 E-3	9.44 E-2	0.08	0
Spinach	0.26	8.39 E-3	0.04	0.48	0.08
Spotted snakehead	0.24	1.03 E-2	4.77 E-2	2.82	8.05 E-2
Tatina	0.27	1.21 E-2	5.35 E-3	0.36	4.35 E-2
MozambiqueTalipia	0.26	9.31 E-3	1.24 E-2	0.28	7.10 E-3

Amaranth and MozambiqueTalipia suggested that this foodstuff might cause human health risk due to higher accumulation of heavy metal in edible portion.

Health risk assessment

The health risk assessment associated with heavy metal (Cu, Co, Fe, Zn, Mn) in locally grown vegetables and fishes of Buriganga River, estimated exposure and risk index were calculated (Table 5). Health risk index was found bellow one (1) in all varieties of vegetables, and fishes for all the measured elements except Zn. In a study Singh et al. (2010) reported Cu, Zn and Cr were not found to cause any risk to the people by consuming vegetables and cereals grown in the area around Dinapur Sewage treatment plant, India. For Zn it was higher in Stem Amaranth and Spotted Snakehead and hence can pose negative impact on human health those who consumes. On the other hand HRI value lower than 1 for other elements in all varieties of vegetables and fishes may be due to take lower amounts in diet, which consequently decreased the health risk index.

Conclusion

Irrigation of agricultural land and fish culturing with treated and untreated wastewater led to the accumulation of heavy metals in vegetables and fishes. Variations in the heavy metal concentrations in the studied vegetables and fishes reflect the difference in uptake capabilities and their further translocation to the edible portion of plants and fishes. The mean concentration of heavy metals in water was lower than the safe limit except Fe and Mn. All the vegetables containing heavy metals lower than recommended tolerable levels proposed by FAO/WHO (2001) except Fe and Co and for fishes it is lower than the permissible level set by WHO (1989) except Mn. The metal pollution index, for all varieties of vegetables and fishes are quite high. Consumption of foodstuff with elevated levels of heavy metals may lead to high level of accumulation in the body causing different disease like thalassemia, dermatitis, brain and kidney damage and cancer.

The level of heavy metals (HMs) found in different sources in the present study was compared with the prescribed safe limit provided by WHO (1989) and (2001), WHO (Pescod 1992), World average value (2000). In the present study fish and vegetable samples from uncontaminated area were not analysed but concentrations of metal in fish and vegetables found are compared with the results of different similar studies carried out all over the world and also with the safe limit suggested by different organization. Among five heavy metals studied in vegetables and fishes, concentration of Fe was maximum. The MPI (metal pollution index) value for Spinach was very high compared to other vegetables

analysed and suggested to take less amount in diet of local people. Health risk index (HIR) for all HMs were less than 1, which is may be due to the high level of allowable oral reference dose of the respective metal and this indicate no risk. The study suggests that as there is high concentration of heavy metals in water, its long term use caused heavy metal contamination leading to health risk of consumers. Thus, regular monitoring of these toxic heavy metals in water, vegetables and fishes is essential to prevent their excessive build up in food chain.

Competing interests
The authors declared that there is no conflict of interest.

Authors' contributions
Sample is collected by FEK. Analysis was carried out by YNJ, FEK, SA and JK. Final manuscript was drafted and edited by YNJ, FEK and GMR. Islam. All authors read and approved the final manuscript.

Acknowledgement
Highly appreciate the association and cooperation of the stuff member of Chemistry division, Atomic energy centre, Dhaka and Department of food Engineering & Tea Technology, Shahjalal University of Science and Technology, Sylhet.

Author details
[1]Department of food Engineering & Tea Technology, Shahjalal University of Science and Technology, Sylhet, Bangladesh. [2]Chemistry Division, Atomic Energy Centre, Dhaka, Bangladesh.

References
Ahmad JU, Goni MA (2010) Heavy metal contamination in water, soil and vegetables of the industrial areas in Dhaka, Bangladesh. Environ Monit Assess 166:347–357

Ahmad MK, Islam S, Rahman S, Haque MR, Islam MM (2010) Heavy metals in water, sediment and some fishes of Buriganga River. Bangladesh Int J Environ Res 4(2):321–332

Alam MGM, Snow ET, Tanaka A (2003) Arsenic and heavy metal concentration of vegetables grown in Samta village, Bangladesh. Sci Total Environ 111:811–815

Al-Nakshabandi GA, Saqqar MM, Shatanawi MR, Faygad M, Al-Horani H (1997) Some environmental problems associated with the use of treated wastewater for irrigation in Jordan. Agric Water Manag 34:81–94

Arora M, Kiran B, Rani A, Rani S, Kaur B, Mittal M (2008) Heavy metal accumulation in vegetables irrigated with water from different sources. Food Chem 111:811–815

Asaduzzaman ATM, Nury SN, Hoque S, Sultana S (2002) Water and soil contamination from tannery waste: potential impact on public health in Hazaribag and surroundings, Dhaka, Bangladesh. Atlas Urban Geol 14:415–444

Azom MR, Mahmud, Yahya KS, Sontu MA, Himon SB (2012) Environmental impact assessment of tanneries: a case study of Hazaribag in Bangladesh. Int J Environ Sci Dev 3(2):152–156

Cui YJ, Zhu YG, Zhai RH, Chen DY, Huang YZ, Qui Y, Liang JZ (2004) Transfer of metals from soil to vegetables in an area near a smelter in Nanning, China. Environ Int 30:785–791

Delbari AS, Kulkarni DK (2013) Determination of heavy metal pollution in vegetables grown along the roadside in Tehran– Iran. Ann Biol Res 4(2):224–233

Eletta OAA, Adekola FA, Omotosho JS (2003) Determination of concentration of heavy metals in two common fish species from ASA River, Ilorin, Nigeria. Toxicol Environ Chem 85(1–3):7–12

FAO/WHO (2001) Food Standards Program. Codex Alimentarius Commission (FAO/WHO) Food additives and contaminants, Geneva, Switzerland, ALINORM 01/12A, pp 1–289

Farooq M, Anwar F, Rashid U (2008) Appraisal of heavy metal contents in different vegetables grown in the vicinity of an industrial area. Pak J Bot 40 (5):2099–2106

Food and Nutritional Board (2004) Dietary Reference Intakes [DRIs]. Recommended Intake for Individuals. National Academy of Sciences, Washington, DC: USA

Friberg L, Nordberg GF, Vpuk B (1984) Handbook on the Toxicity of Metals. Elsevier, Nort Holland, Bio Medical Press, Amsterbam

Gebrekidan A, Weldegebriel Y, Hadera A, Vander Bruggen B (2013) Toxicological assessment of heavy metals accumulated in vegetables and fruits grown in Ginfrel River near Sheba tannery, Tigray, Northern Ethiopia. Ecotox Environ Safety 95:171–178

Gupta N, Khan DK, Santra SC (2008) an assessment of heavy metal contamination in vegetables grown in wastewater-irrigated areas of Titagarh, West Bengle, India. B Environ Contam Tox 80:115–118

Hasan N, Ahmad K (2000) Intra-familial distribution of food in rural Bangladesh. Institute of Nutrition and Food Science. University of Dhaka, Bangladesh

Hina B, Rizwani GH, Naseem S (2011) Determination of toxic metals in some herbal drugs through atomic absorption spectroscopy. PakJPharmSci 24 (3):353–358

Hossain MAR, Nahiduzzaman M, Sayeed MA, Saha D, Azim ME (2008) Fish consumption amongst poor people in Bangladesh: Effects of gender, location and wealth class. Aquaculture News, Bangladesh

Industrial Effluent Quality Criteria (1999) A comparision of Environmental Laws, Bangladesh Gazzette Additional 28, department of Environment (DoE). The Ministry of Environment and Forests, Bangladesh, p 60

Jan FA, Ishaq M, Khan S, Shakirullah M, Asim SM, Ahmad I (2011) Bioaccumulation of metals in human blood in industrially contaminated area. J Environ Sci 23(12):2069–2077

Jolly YN, Islam A, Akbar S (2013) Transfer of metals from soil to vegetables and possible health health risk assessment. Springerplus 2:385. doi:10,1186/2193-1801-2-385

Kennedy G, Burlingame B, Nguyen VN (2001) Nutritional contribution of rice and impact of biotechnology and biodiversity in rice-consuming countries. Crop and Grassland Service, FAO, Rome, Italy

Khan YSA, Hossain MS, Hossain SMGMA, Halimuzzaman AHM (1998) An environment of trace metals in the GMB Estuary. J Remote Sensing Environ 2:103–113

Khan S, Cao Q, Zheng YM, Huang YZ, Zhu YG (2008) Health risks of heavy metals in contaminated soils and food crops irrigated with wastewater in Beijing, China. Environ Pollut 152:686–692

Liu WH, Zhao JZ, Ouyang ZY, Soderlund L, Liu GH (2005) Impacts of sewage irrigation on heavy metals distribution and contamination. Environ Int 31:805–812

Mingorance MD, Valdes B, Oliva Rossini S (2007) Strategies of heavy metal uptake by plants growing under industrial emissions. Environ Int 33(4):514–520

Mokhtar MB, Aris AZ, Munusamy V, Praveena SM (2009) Assessment level of heavy metals in Penaeus Monodon and Oreochromis Spp in selected aquaculture ponds of high densities development area. Eur J Sci Res 30 (3):348–360

Muchuweti M, Birkett JW, Chinyanga E, Zvauya R, Scrimshaw MD, Lester J (2006) Heavy metal content of vegetables irrigated with mixture of wastewater and sewage sludge in Zimbabwe: Implications for human health. Agric Ecosyst Environ 112:41–48

Muiruri JM, Nyambaka HN, Nawiri MP (2013) Heavy metals in water and tilapia fish from Athi-Galana-Sabaki tributaries, Kenya. Int Food Res J 20(2):891–896

Muwanga A, Barifaijo E (2006) Impact of industrial activities on heavy metal loading and their physic-chemical effects on wetlands of lake Victoria basin (UGANDA). Afr J Sci Technol 7(1):51–63

Naser HM, Sultana S, Mahmud NU, Gomes R, Noor S (2011) Heavy metal levels in vegetables with growth stage and plant species variations. Bangladesh J Agril Res 36(4):563–574

Okunola OJ, Alhassan Y, Yapbella GG, Uzairu A, Tsafe AI, Abechi ES, Apene E (2011) Risk assessment of using mobile phone recharge cards in Nigeria. J Environ Chemistry Ecotox 3(4):80–85

Pendias AK, Pendias H (2000) Trace elements in Soils and Plants. CRC press, FL, United States, pp 10–11

Pescod MB (1992) Wastewater treatment and use in agriculture.FAO irrigation and Drainage paper 47. Food and Agriculture Organization of United Nations, Rome

Saeed S, Shaker MI (2008) Assessment of heavy metals pollution in water and sediments and their effect on Oreochromis niloticus in the Northern Delta Lakes, Egypt. Int Symposium on Tilapia Aquaculture 8:475–490

Sharma RK, Agrawal M, Marshall F (2006) Heavy metal contamination in vegetables grown in wastewater irrigated areas of Varanasi, India. B Environ Contam Tox 77:312–318

Sharma R, Agrawal M, Marshall F (2007) Heavy metal contamination of Soil and Vegetables in suburban areas of Varanasi, India. Ecotoxicol Environ Saf 66:258–266

Singh A, Sharma RK, Agarwal M, Narshal FM (2010) Health risk assessment of heavy metal via dietary intake of foodstuffs from the wastewater irrigated site of a dry tropical area of India. Food Chem Toxicol 48:611–619

Tandi NK, Nyamangara J, Bangira C (2004) Environmental and potential health effects of growing leafy vegetables on soil irrigated using sewage sludge and effluent: a case of Zn and Cu. J Environ Sci Health B 39:461–471

Tripathi RM, Raghunath R, Krishnamoorthy TM (1997) Dietary intake of heavy metals in Bombay City, India. Sci Total Environ 208:149–159

Ubalua AO, Chijioke UC, Ezeronye OU (2007) Determination and assessment of heavy metal content in fish and shellfish in Aba River, Abia State, Nigeria. KMITL Sci Tech J 7(1):16–23

Ureso J, Gonzalez-Regalado E, Gracia I (1997) Trace elements in bivalvemollusks Ruditapes decussates and Ruditapes phillippinarum from Atlantic Coast of Southern Spain. Environ Int 23(3):291–298

US Environmental Protection Agency (US EPA) (1989) Risk Assessment Guidance for Superfund: Human Health Evaluation Manual [part A]: Interim Final. U.S. Environmental Protection agency, Washington, DC, USA [EPA/540/1-89/002]

US Environmental Protection Agency (US EPA) (2002) Region 9, Prelominary Remidation Goals

WHO (1989) Heavy metals-environmental aspects. Environment Health Criteria. No. 85, Geneva, Switzerland

WHO (1993) Evaluation of certain food additives and contaminants in Forty-First report of the joint FAO/WHO expert committee on food additives. WHO, Geneva, Switzerland, WHO technical series, 837

Yoon KP (2003) Construction and characterization of multiple heavy metal-resistant Phenol-degrading pseudomonads strains. J Micro Biotech 13 (6):1001–1007

Zhuang P, McBride BB, Xia HP, Liny, Liza (2009) Health risk from heavy metals viz consumption of food crops in vicinity of Dabaoshan mine, South China. Sci Total Environ 407:1551–1561

2

Escherichia coli O157:H7 in Raw Meat in Addis Ababa, Ethiopia: Prevalence at an Abattoir and Retailers and Antimicrobial Susceptibility

Tizeta Bekele[1], Girma Zewde[1], Genene Tefera[2], Aklilu Feleke[1] and Kaleab Zerom[3*]

Abstract

Background: Although raw meat and its products are commonly consumed in traditional Ethiopian diets, *E. coli* O157: H7 is rarely studied compared to other countries. Thus the present study has been designed to determine the prevalence and assess the antimicrobial susceptibility of *E. coli* O157: H7 isolated from beef, sheep meat and goat meat at one abattoir and in 48 selected raw meat retail shops in Addis Ababa.

Results: Out of 384 meat samples examined, 10.2% (39/384) were positive to *E. coli* O157:H7. Among these samples examined, beef was the most frequently contaminated with *E. coli* O157:H7 with an overall prevalence of 13.3% (17/128) followed by 9.4% (12/128) sheep meat and 7.8% (10/128) goat meat. With regard to meat source, the prevalence rates of *E. coli* O157:H7 at the abattoir and the selected retail shops were 5.7% (11/192) and 14.6% (28/192), respectively. Significant differences in prevalence was observed among sample sources (p < 0.05). The antimicrobial susceptibility investigation of 39 *E. coli* O157:H7 isolates using 10 commercially available antimicrobial discs revealed that the isolates were susceptible to nine antimicrobials from 69.3% to 100% except streptomycin which showed susceptibility of 48.7%. An overall resistance of 33.4% and 30.9% was recorded to streptomycin and amikacin, whereas 5.1%, 5.1%, 7.7%, 12.8% and 17.9% resistance rates were recorded against nalidixic acid, tetracycline, amoxacilin-clavulanic acid, cephalothin and ciprofloxacin, respectively. Multidrug resistance was observed among amikacin, amoxycillin-clavulanic acid, cephalothin, ciprofloxacin, streptomycin and tetracycline antimicrobials drugs.

Conclusions: The isolation of *E. coli* O157:H7 in raw meat and the existence of antimicrobial resistant isolates highlight the potential threat to public health. Hence implementation of *E. coli* O157:H7 prevention and control strategies from farm production to consumption of meat and meat products are crucial.

Keywords: Abattoir; Addis Ababa; Antimicrobials; *E. coli* O157:H7; Prevalence; Retail shops; Raw meat

Background

Most *Escherichia coli* (*E. coli*) are normal commensals found in the intestinal tract of both humans and animals, while others are pathogenic to humans (CFSPH 2009; Kaper et al. 2004). Pathogenic *E. coli* distinguished from normal flora by their possession of virulence factors. The specific virulence factors can be used, together with the type of disease, to separate these organisms into pathotypes (CFSPH 2009). *E. coli* O157: H7 is one of the best known serotype to contain pathotypes that can cause food borne infection in humans (Acha and Szyfres 2001; IFT 2003). *E. coli* O157:H7 pathotypes has been found in the intestines of healthy cattle, deer, goats, and sheep (Acha and Szyfres 2001; IFT 2003).

Verocytotoxigenic (or verotoxigenic) *E. coli* (VTEC) pathotype produces a toxin known as verocytotoxin that is lethal to cultured African green monkey kidney cells (Vero cells) (CFSPH 2009). VTEC may belong to many serotypes, but most severe human infections are caused by *E. coli* O157: H7 serotypes (Mead and Griffin 1998). Enterohemorrhagic *E. coli* O157: H7 (EHEC) is VTEC that possess additional virulence factors, giving them the ability to cause hemorrhagic colitis and hemolytic uremic syndrome (HUS) in humans. EHEC have been described as important food

* Correspondence: kaleabzerom@yahoo.com
[3]College of Veterinary Medicine, Samara University, P.O. Box 132, Samara, Ethiopia
Full list of author information is available at the end of the article

borne pathogens (Vernozy-Rozand 1999). Young children, the elderly, and those with weakened immune systems are the most vulnerable for infection (IFT 2003).

The most frequent mode of transmission for *E. coli* O157:H7 infection to human is through consumption of contaminated food and water. However, it may also spread directly from person to person and occasionally through occupational exposure (Abdalla et al. 2009).

Ruminants, particularly cattle and sheep, are the most important reservoir hosts for EHEC O157:H7 (Cornick et al. 2000). Cattle have been identified as a major reservoir of *E. coli* O157 (Chapman et al. 2001) and consumption of foods of bovine origin have been associated with some of the largest food poisoning outbreaks in which this organism was identified as the etiologic agent (Meng and Doyle 1998).

The microbiological contamination of carcasses can occur during processing and manipulation, such as skinning, evisceration, storage and distribution at slaughter houses. Fecal matter is a major source of contamination and could reach carcasses through direct deposition, as well as by indirect contact through contaminated equipment, workers, installations and air (Pal 2007). Cattle slaughtering operations, such as bleeding, dressing and evisceration expose sterile muscle to microbiological contaminants that were present on the skin, the digestive tract and in the environment (Bacon et al. 2000; Abdalla et al. 2009).

There are a few reports on the prevalence and antibiotic resistance status of *E. coli* O157:H7 from raw meat in Ethiopia and little or no studies have been carried out in Addis Ababa. Thus the current study has been designed to determine the prevalence and assess the antimicrobial susceptibility of the microorganism isolated from beef, sheep and goat's meat at one abattoir and in selected meat retail shops in Addis Ababa. The difference between the abattoir included in this study and in the study conducted by Hiko et al. (2008) with respect to the level of establishment (the abattoir included in this study supply meat for local consumption whereas the abattoir in Hiko et al. (2008). study is an export abattoir that fulfill most of international standards), the slaughtering operation and the food safety management system may enable this study to build further on the data provided by Hiko et al. (2008).

Methods
Study area
The study was carried out in Addis Ababa, the capital city of the Federal Democratic Republic of Ethiopia.

Description of food business operators included in the study
The food business operators included in this study, both the abattoir and the selected 48 retail meat shops, do not use any formal food safety management system. The abattoir supplies meat to all of the retailers studied. All of the retail meat shops included in this study does not have any chilling/cooling facilities.

Study design
A cross-sectional study was conducted to determine the prevalence and assess the antimicrobial susceptibility of *E. coli* O157: H7 isolated from beef, sheep and goat meat at one abattoir and in 48 selected raw meat retail shops from August 2011 to April 2012 in Addis Ababa.

Study samples
The study was conducted on a total of 384 raw meat samples consisting of 64 beef, 64 sheep meat and 64 goat meat each collected from one abattoir (n = 192) and 48 randomly selected retail shops (n = 192) that means with the total of 128 beef, 128 sheep meat and 128 goat meat.

Sampling technique and sample collection procedure
Swab samples were collected aseptically using systematic random sampling technique from the carcasses of daily slaughtered animals at the abattoir and from legally registered raw meat retailer shops using simple random sampling technique.

Selected carcasses were swabbed using the method described in ISO17604 (2005) by placing sterile template (10 x 10 cm) on specific sites of a carcass. A sterile cotton tipped swab, (2X3 cm) fitted with shaft, was first soaked in an approximately 10 ml of buffered peptone water (Oxoid Ltd., Hampshire, England) rubbed first horizontally and then vertically several times on the carcasses. The abdomen (flank), thorax (lateral) and breast (lateral); which are sites with the highest rate of contamination (ISO 17604, 2005); were chosen for sampling using the same swab over all the sites. On completion of the rubbing process, the shaft was broken by pressing it against the inner wall of the test tube and disposed leaving the cotton swab in the test tube. 25 g of whole cuts of raw meat sample was collected from retail shops following aseptic techniques. The samples were put in a sterile universal bottle filled with 225 ml of buffered peptone water. Finally, the samples were transported to the Institute of Biodiversity Conservation Laboratory, Addis Ababa using ice box in cold chain for microbiological analysis. Up on arrival, the samples were stored in refrigerator at 4°C for 24 hrs until being processed for isolation.

Sample preparation
Each carcass swabs were homogenized with vortex mixer and 25 g of raw whole cuts of meat sample collected from each retail shops were taken out from the universal

bottle, chopped aseptically and the meat was placed with 225 ml of buffered peptone water; from the previous universal bottle; in a plastic bag and homogenized using a homogenizer (Stomacher 400, Seward Medical, England) at high speed for 2 minutes. The resulting suspension was used for isolation of *E. coli* O157:H7.

Isolation Procedure

200 μl of prepared sample of a meat rinse solution was streaked onto MacConkey agar (IVD, UK) plates and incubated at 37°C for 24 hrs. Following incubation, lactose-positive (pink) colonies were streaked onto Sorbitol-MacConkey Agar (IVD, UK) plates and incubated at 37°C for 24 hrs. The sorbitol negative colorless colonies were sub-cultured on Rainbow agar.

Rainbow Agar O157 (Hayward, USA) was inoculated by spreading a suspected colonies of *E. coli* on its surface. The plates were then incubated for 20 to 24 hrs, or longer, at 37°C and observed for the presence of colored colonies. The distinctive black or gray coloration of *E. coli* O157 colonies were easily viewed by laying the Petri dish against a white background. Upon sub-culturing, the isolated *E. coli* O157 colonies showed their typical black or gray coloration (BiOLOG User Guide 2008). These colonies were tested further using BiOLOG identification system to confirm their identity as *E. coli* and for any isolate identified as *E. coli* O157:H7 additional confirmation by serology may be recommended for the reasons of serious implications when this pathogen is determined in a food sample.

Identification

Pure colonies that showed typical black or gray coloration on Rainbow agar were inoculated on BUG (Biolog Universal Growth Medium) agar (Hayward, USA) with 5% sheep blood and incubated at 37°C for 24 hrs. Subculture was made using the same culture media to have pure culture colonies before identification was done by OmniLog. The BUG is a recommended medium for aerobic bacteria and it was employed to isolate *E. coli*. Following this, the identification of *E. coli* O157:H7 was performed using the Omnilog plus Identification System (BiOLOG User Guide 2008).

After getting pure culture colonies, identification was carried out using BiOLOG Standard Operation Protocols (SOP). Briefly, bacterial suspension was prepared with an appropriate level of bacterial density, as recommended in the protocol of the instrument. The bacterial suspension was inoculated into the GEN III Micro Plates aseptically. The microplates were covered with lid and incubated at 37°C for 22 hrs. Then the Micro Plates were loaded into the OmniLog incubator/reader. The bacterial suspension was identified by the instrument using the inbuilt database.

Antimicrobial susceptibility pattern

The antimicrobial susceptibility test was performed following the standard agar disk diffusion method according to CLSI (2008) using commercially available antimicrobial disks (Table 1).

Table 1. Antibiotic disks used to test *E. coli* O157:H7 and their respective concentrations.

Each isolated bacterial colony from pure fresh culture was transferred into a test tube of 5 ml Tryptone Soya Broth (TSB) (Oxid, England) and incubated at 37°C for 6 hrs. The turbidity of the culture broth was adjusted using sterile saline solution or added more isolated colonies to obtain turbidity usually comparable with that of 0.5 McFarland standards (approximately 3×10^8 CFU per ml). Mueller-Hinton agar (Bacton Dickinson and Company, Cockeysville USA) plates were prepared according to the manufacturer guidelines. A sterile cotton swab was immersed into the suspension and rotated against the side of the tube to remove the excess fluid and then

Table 1 Antibiotic disks used to test *E. coli* O157:H7 and their respective concentrations

Nº	Antibiotic disks	Disc code	Concentration	Diameter of zone of inhibition in millimeter (mm)		
				Resistant ≤	Intermediate	Susceptible ≥
1	Amikacin	AK	30 μg	14	15-16	17
2	Amoxycillin-Clavulanic acid	AMC	20/10 μg	13	14-17	18
3	Ceftriaxone	CRO	30 μg	14	15-17	18
4	Cephalothin	KF	30 μg	14	15-17	18
5	Chloramphenicol	C	30 μg	12	13-17	18
6	Ciprofloxacin	CIP	5 μg	15	16-20	21
7	Nalidixic acid	NA	30 μg	13	14-18	19
8	Streptomycin	S	10 μg	11	12-14	15
9	Tetracycline	TE	30 μg	11	12-14	15
10	Sulfamethoxazole - Trimethoprim	SXT	30 μg	10	11-15	16

swabbed in three directions uniformly on the surface of Mueller-Hinton agar plates. After the plates dried, antibiotic disks were placed on the inoculated plates using sterile forceps. The antibiotic disks were gently pressed onto the agar to ensure firm contact with the agar surface, and incubated at 37°C for 24 hrs. Following this the diameter of inhibition zone formed around each disk was measured using a black surface, reflected light and transparent ruler by lying it over the plates. The results were classified as sensitive, intermediate, and resistant according to the standardized table supplied by the manufacturer (CLSI 2008). *E. coli* ATCC 25922 type strains was used as a positive control.

Data management and analysis

The coded data was entered in MS Excel (Additional file 1) and then analyzed using SPSS version 15 (2006). The overall prevalence of *E. coli* O157: H7 in raw meat was determined using standard formula. The number of positive samples was divided by the total number of samples examined multiplied by 100. In addition to these, the prevalence in each meat type as well as at the selected retail shops and the abattoir was determined in the same way by dividing positive value with corresponding total examined samples. Difference among and between proportions of the groups with certain determinant factor was determined by chi-square (χ^2) test. A p-value <0.05 was considered indicative of a statistical significant difference.

Description of raw data

The following raw data are available with the online version of this paper with sample identification and corresponding results. The data file 1 (Additional file 1) is a table listing the type of meat (beef, sheep meat, goat meat) with their corresponding result; *E. coli* O157:H7 positive meat samples were represented with 1 whereas negative samples as 0. Data file 2 (Additional file 1) contains raw data for antimicrobial susceptibility of *E. coli* O157: H7 isolated from beef, sheep meat and goat meat at an abattoir and in 48 selected raw meat retail shops in Addis Ababa.

Results

Prevalence

The overall prevalence of *E. coli* O157:H7 in three types of raw meat samples (beef, sheep and goat's meat) was 10.2% (39/384). Out of which 13.3% (17/128) were from beef, 9.4% (12/128) from sheep meat and 7.8% (10/128) from goat meat. The test statistics among three types of raw meat samples indicated that there was no statistical significance difference in prevalence rate (p > 0.05) (Table 2).

Table 2. Prevalence of *E. coli* O157:H7 in abattoir and retail shops from different type of meat samples.

Among the food business operators, higher prevalence of *E. coli* O157:H7 was found in the retailer shops (14.6%) than the abattoir (6.3%) and there is significant difference in prevalence (p < 0.05) (Table 2). However there was no statistically significant relationship difference in prevalence of *E. coli* O157:H7 isolated from the three types of meat both in the abattoir and in the retailer shops (Table 2).

Antimicrobial susceptibility pattern

The result of antimicrobial susceptibility test of 39 *E. coli* O157:H7 isolated from raw meat samples with 10 selected antimicrobial agents is shown in Table 3. The antimicrobial sensitivity test of *E. coli* O157:H7 isolated from different raw meat types revealed a varying degree of susceptibility to antimicrobial agents tested (Table 3).

Isolates recovered from sheep meat were found to be 40% to 80% susceptible to five antimicrobial agents tested except amoxicillin-clavulanic acid, ceftriaxone, cephalothin, chloramphenicol, and sulfamethoxazole-trimethoprim antimicrobial agents, which showed 100% susceptibility (Table 3). Similarly, *E. coli* O157:H7 isolates from goat meat showed 100% susceptiblity to amikacin, ceftriaxone, chloramphenicol, sulfamethoxazole-trimethoprim and tetracycline. However, the remaining isolates showed a susceptibility ranging from 58.3% to 91.7%. *E. coli* O157:H7

Table 2 Prevalence of *Escherichia coli* O157:H7 in abattoir and retail shops from different type of meat samples

Food.business operators	Samples examined	№ (%) examined	№ (%) + ve	χ^2	P-value
	Beef	64(33.3)	3(4.7)		
Abattoir	Sheep meat	64(33.3)	4(6.3)	0.193	0.908
	Goat meat	64(33.3)	4(6.3)		
Total		**192(50)***	**11(5.7)***		
	Beef	64(33.3)	14(21.9)		
Retail shops	Sheep meat	64(33.3)	8(10.9)	4.348	0.114
	Goat meat	64(33.3)	6(9.4)		
Total		**192(50)***	**28(14.6)***		

+ ve = E coli contaminated; p-value between food business operators (abattoir and retail shops)*= 0.04.

recovered from beef showed a susceptibility to antibiotic ranging from 17.6% to 100%.

Table 3. Antimicrobial susceptibility pattern of *E. coli* O157:H7 isolates (n = 39) to10 selected antimicrobial agents.

Out of 39 *E. coli* O157:H7 isolates subjected to antimicrobial susceptibility test a total of 23.1% were susceptible to all antimicrobials used; from these 55.5% isolates were from beef, 22.2% from sheep meat and goat meat each (Table 4).

Table 4. *E. coli* O157:H7 isolates susceptible to all antimicrobials.

Of the 39 *E. coli* O157:H7 isolates, 17.9% were found to be resistant to three or more drugs tested of which 40% were recovered from beef samples (Table 5).

Table 5. Multidrug resistance (MDR) of *E. coli* O157:H7.

Discussion

Prevalence of E. coli O157:H7

Raw meat and its products are commonly consumed in traditional Ethiopian diets, but *E. coli* O157:H7 is rarely studied compared to other countries. In the present study, *E. coli* O157:H7 was isolated from beef, sheep meat and goat meat at both the abattoir and the selected raw meat retailer shops at Addis Ababa. The overall prevalence of *E. coli* O157:H7 in the present study (10.2%) was higher than the 4.2% reported in Modjo and Bishftu (Debre Zeit) (Hiko et al. 2008) in Ethiopia using immuno-magnetic separation method which is much more sensitive than the plating method used in the present study.

The difference in the overall prevalence observed among the three types of meat samples in present study is high (13.3%) in beef, but relatively similar between sheep meat (9.4%) and goat meat (7.8%). In contrast, the prevalence of *E. coli* O157:H7 in beef in this study was higher than the 8.8% prevalence reported by Abong (2008) in South Africa and Hajian et al. (2011) in Iran; and lower than 53% prevalence reported by Dahiru et al. (2008) in fresh beef meat in Nigeria.

In this study, the prevalence of *E. coli* O157:H7 in sheep meat (9.4%) and goat meat (7.8%) were higher than the previous study done by Hiko et al. (2008) in Ethiopia who reported a prevalence of 2.5% in sheep meat and 2% in goat meat. The presence of *E. coli* O157: H7 in sheep and goat meat might be due to contamination either from gastrointestinal content and/or skin (McEvoy et al. 2004).

With regard to meat source the higher prevalence of *E. coli* O157:H7 was found at the retailer shops (14.6%) than the abattoir (5.7%). The significant variation (p < 0.05) in prevalence (contamination) rate between the abattoir and the retailer shops could be due to the difference in hygienic practices, cooling and storage time used between the two. Moreover there could be risk of carcass contamination and cross and subsequent contamination, during transportation, environment, handling of meat at retailer shops.

The prevalence of *E. coli* O157:H7 was similar at the abattoir in beef, sheep meat and goat meat. There is no significant variation (p > 0.05) among the three type of raw meat samples. However difference in prevalence was observed among the three types of meat samples from raw meat from the retailer shops. The high prevalence was recorded in beef (21.9%) than sheep meat (10.9%) and goat meat (9.4%). This could be due to the fact that bovine has been implicated as the principal reservoir of this pathogen as compared with other food animals (Fantelli and Stephan 2001).

Table 3 Antimicrobial susceptibility pattern of *E. coli* O157:H7 isolates (n = 39) to10 selected antimicrobial agents

Antimicrobial used	Type of raw meat and *E. coli* O157:H7 isolates											
	Beef (n = 17)			Sheep meat (n = 12)			Goat meat (n = 10)			Total (n = 39)		
	S	I	R	S	I	R	S	I	R	S	I	R
	Nº (%)	Nº (%)	Nº (%)	Nº (%)	Nº (%)	Nº (%)	Nº (%)	Nº (%)	Nº (%)	Nº (%)	Nº (%)	Nº (%)
AK	10(58.8)	0(0)	7(41.2)	10(83.6)	0(0)	2(16.6)	7(70)	0(0)	3(30)	27(69.3)	0(0)	12(30.9)
AMC	17(100)	0(0)	0(0)	9(75)	0(0)	3(25)	10(100)	0(0)	0(0)	36(92.3)	0(0)	3(7.7)
CRO	17(100)	0(0)	0(0)	12(100)	0(0)	0(0)	6(60)	4(40)	0(0)	39(100)	0(0)	0(0)
KF	15(88.2)	2(11.8)	0(0)	7(58.3)	0(0)	5(41.7)	10(100)	0(0)	0(0)	32(82.1)	2(5.1)	5(12.8)
C	17(100)	0(0)	0(0)	12(100)	0(0)	0(0)	10(100)	0(0)	0(0)	39(100)	0(0)	0(0)
CIP	13(76.5)	0(0)	4(23.5)	9(75)	0(0)	3(25)	6(60)	4(40)	0(0)	28(71.8)	4(10.3)	7(17.9)
NA	17(100)	0(0)	0(0)	10(83.6)	2(16.4)	0(0)	4(40)	4(40)	2(20)	31(79.5)	6(15.4)	2(5.1)
S	4(23.5)	7(4.1)	6(35.3)	11(91.6)	0(0)	1(8.4)	4(40)	0(0)	6(60)	19(48.7)	7(17.9)	13(33.4)
TE	17(100)	0(0)	0(0)	12(100)	0(0)	0(0)	8(80)	2(20)	2(20)	37(94.9)	0(0)	2(5.1)
SXT	17(100)	0(0)	0(0)	12(100)	0(0)	0(0)	10(100)	0(0)	0(0)	39(100)	0(0)	0(0)

See Table 2 for key abbreviations; S = Sensitive, I = Intermediate, R = Resistant.

Table 4 *E. coli* O157:H7 isolates susceptible to all antimicrobials

Type of raw meat	Nº of isolates tested	Nº (%) of isolates susceptible to all antimicrobials
Beef	17	5 (55.5)
Sheep meat	12	2 (22.2)
Goat meat	10	2 (22.2)
Total	39	9 (23.1)

Antimicrobial susceptibility pattern of E. coli O157:H7

The use of antibiotics in the treatment of infection with *E. coli* O157:H7 is controversial, since antimicrobial therapy may increase the risk of development of HUS (Hemolytic uremic syndromes) (Mølbak et al. 2002). Although some studies do not advice antibiotic treatment for infections caused by such bacteria (Wong et al. 2000), others suggest that disease progression may be prevented by administrating antibiotic at early stage of infection (Shiomi et al. 1999). Thus, for better response, antimicrobial susceptibility test is necessary (Quinn et al. 2002). Hence, on the basis of this necessity, antimicrobial susceptibility test was conducted on the isolates recovered from raw meat.

Antimicrobial resistance of *E. coli* O157:H7 isolates from animal and human sources have been reported in Ethiopia by Hiko et al. (2008). In the present study *E. coli* O157:H7 showed resistance to seven antimicrobials which varied from 5.1% to 33.4% except to ceftriaxone, chloramphenicol, and sulfamethoxazole-trimethoprim to which 100% susceptibility was noticed. The 100% susceptibility of all the isolates to chloramphenicol, ceftriaxone and sulfamethoxazole-trimethoprim is consistent with the findings of Rangel and Marin (2009) and Rahimi and Nayebpour (2012). Most of these antimicrobials are not commonly used in Ethiopia in the treatment of animals that served as a source of meat. Moreover, the susceptibility might have contributed to the effectiveness of these antimicrobials mostly against Gram negative bacteria like those of the family of Enterobacteriaceae to which *E. coli* O157:H7 belongs.

Table 5 Multidrug resistance (MDR) of *E. coli* O157:H7

Type of drugs registered as MDR	Source of resistant isolates			Total Nº (%)
	Beef (Nº)	Sheep meat (Nº)	Goat meat (Nº)	
AK, S, TE	-	-	2	2 (28.6)
AK, CIP, S	2	-	-	2 (28.6)
AMC, CIP, S	1	-	-	1 (14.3)
AK, KF, CIP, S	1	-	-	1 (14.3)
AK, AMC, KF, S	-	1	-	1 (14.3)
Total MDR Nº (%)	4 (57.1)	1 (14.3)	2 (28.6)	7 (17.9)

See Table 2 for key abbreviations; S = Sensitive, I = Intermediate, R = Resistant.

The high resistance to streptomycin in this study is in agreement with Hiko et al. (2008) who reported antimicrobial resistance to *E. coli* O157:H7 isolates from raw meat samples to some of above mentioned antimicrobials especially to streptomycin. The significantly high level of resistance to these antimicrobials was probably an indication of their extensive usage in the veterinary sector for therapeutic and prophylactic purpose both for *E. coli* and other infections.

Antimicrobial resistance emerges from the use of antimicrobials in animals and human, and the subsequent transfer of resistance genes and bacteria among animals, humans and animal products and the environment (Scott et al. 2002). The shedding of the resistant bacteria into the environment by cattle may lead to a widespread dissemination of antibiotic resistant genes to the resident bacteria in the environment (Callaway et al. 2003; Mashood, et al. 2006). Evidence has been found which indicates that resistant strains of pathogens can be transmitted to humans through food (Oosterom 1991; Khachatourians 1998). Antibiotic resistance among foodborne pathogens may create an increased burden to human health through: it's potential to reach humans, increasing the risk of acquiring an infection in human who taking prior antibiotic treatment, limiting illness treatment options and may be by developing increased virulence (IFT 2006).

Recently, multidrug resistant (MDR) phenotypes have been spread widely among Gram negative bacteria (Ahemed et al. 2006). MDR was observed among amikacin, amoxycillin-clavulanic acid, cephalothin, ciprofloxacin, streptomycin and tetracycline antimicrobials in this study. This is in agreement with the findings by other researchers, who reported multidrug resistance among *E. coli* O157:H7 isolates (Kim et al., 1994; Schroeder et al., 2002). From the above mentioned antimicrobials streptomycin is found in all MDR *E. coli* O157:H7 isolates. This finding was supported by Hiko et al. (2008).

Increased sensitivity of antimicrobial resistant *E. coli* O157 to environmental or food processing related stresses (acid and heat) have been reported by different authors. Duffy et al. (2006) reported that MDR *E. coli* O157:H7 (resistant to 10 antibiotics) when subjected to food stresses (acid and heat) was found to act very differently to the unstressed antibiotic sensitive and antibiotic resistant VTEC strains. All VTEC strains tested were found to survive for approximately 30 days in orange juice at pH 4.4 and 25 days in yoghurt at pH 4.2. The exception was the MDR *E. coli* O157:H7 isolate which was found to have died off significantly faster (P < 0.05) in both media, than in the other strains tested. Thermal inactivation studies also showed the MDR strain to be significantly more heat sensitive (D55 value = 1.71 min) than all other VTEC strains examined (Clavero et al. 1998; Byrne et al. 2002; Huang and Juneja 2003).

Conclusions

One of the most significant food-borne pathogens that have gained increased attention in recent years is *E. coli* O157:H7. In this study, isolation of *E. coli* O157:H7 from raw beef, sheep meat and goat meat and the existence of resistant isolates highlight the potential threat to public health. Hence implementation of *E. coli* O157:H7 prevention and control strategies from farm production to consumption of meat and meat products are crucial.

Additional file

Additional file 1: Raw Data for Escherichia coli O157:H7 in Raw Meat in Addis and Antimicrobial Susceptibility.

Competing interests

The authors declare that they have no competing interests.

Authors' contributions

TB: conception and design, acquisition of data, analysis and interpretation of data, drafting the manuscript; GZ: conception and design, acquisition of data; drafting manuscript; GT: conception and design, acquisition of data laboratory work; AF: acquisition of data, analysis and interpretation of data, drafting the manuscript; KZ: participated in its design, analysis and interpretation of data, revising it critically for important intellectual content. All authors read and approved the final manuscript.

Acknowledgements

The authors are very grateful to Institute of Biodiversity Conservation Microbial Genetic Resources Centre and College of Veterinary Medicine and Agriculture, Addis Ababa University for their financial and laboratory facility provision for running this project.

Author details

[1]College of Veterinary Medicine and Agriculture, Addis Ababa University, P.O. Box 34, Bishftu, Ethiopia. [2]Institute of Biodiversity Conservation, Microbial Genetic Resources Centre, P.O. Box 30726, Addis Ababa, Ethiopia. [3]College of Veterinary Medicine, Samara University, P.O. Box 132, Samara, Ethiopia.

References

Abdalla MA, Siham E, Suliman YYH, Alian A (2009) Microbial Contamination of sheep carcasses at El Kadero slaughter house Khartoum State, Sudan. J Vet Sci 48:1–2

Abong BO (2008) Prevalence of *Escherichia coli* O157:H7 in water and meat and meat products and vegetables sold in the Eastern Cape Province of South Africa and its impact on the diarrheic conditions of HIV/AIDS patients. Dissertation. The University of Fort Hare. Cape Town, South Africa, p 261

Acha PN, Szyfres B (2001) Colibacilosis. In: Zoonoses and Communicable Diseases Common to Man and Animals, 3rd edn. pp 90, Washington, D.C, p 106

Ahemed A, Li J, Shiloach Y, Robbins JB, Szu SC (2006) Safety and immunogenicity of *Escherichia coli* O157 O-specific polysaccharide conjugate vaccine in 2–5 year old children. J Infect Dis 193(4):515–521

Bacon RT, Belk KE, Sofos JN, Clayton RP, Reagan JO, Smith GC (2000) Microbial populations on animal hides and beef carcasses at different stages of slaughter in plants employing multiple sequential interventions for decontamination. J Food Prot 63(8):1080–1086

BiOLOG User Guide (2008) Data Collection Software Identification System User Guide. Biolog, Inc., Hayward, CA

Byrne CM, Bolton DJ, Sheridan JJ, Blair IS, McDowell DA (2002) The effect of commercial production and product formulation stresses on the heat resistance of Escherichia coli O157: H7 (NCTC 12900) in beef burgers. Int J Food Microbiol 79:183–192

Callaway RT, Anderson CR, Elder RO, Edrington ST, Genovese JK, Bischoff MK, Poole LT, Jung SY, Harvey BR, Nisbet JD (2003) Preslaughter Intervention Strategies to Reduce Food-Borne Pathogens in Food Animals. J Anim Sci 81:2

Chapman PA, Cerda'n Malo AT, Ellin M, Ashton R, Harkin M (2001) *Escherichia coli* O157 in cattle and sheep at slaughter, on beef and lamb carcasses and in raw beef and lamb products in South Yorkshire,UK. Int J Food Microbiol 64 (1–2):139–150

Clavero MRS, Beuchat LR, Doyle MP (1998) Thermal Inactivation of Escherichia coli O157: H7 isolated from ground beef and bovine feces, and suitability of media for enumeration. J Food Prot 61:285–289

CLSI (2008) Performance Standards for Antimicrobial Disk and Dilution Susceptibility Tests for Bacterial Isolated from Animals, 3rd edn. CLSI, Wayne, PA, USA, M31-A-3, 28

Cornick NA, Booher SL, Casey TA, Moon HW (2000) Persistent colonization of sheep by *Escherichia coli* O157:H7 and other *E. coli* pathotypes. Appl Environ Microbiol 66(11):4926–4934

Dahiru M, Uraih N, Enabulele SA, Shamsudeen U (2008) Prevalence of *Escherichia Coli* O157:H7 in fresh and roasted beef in Kano City. Nigeria Bayero J Pure Appl Sci 1:39–42

Doyle MP, Institute of Food Technologists (IFT) (2006) Expert Panel Report Summary on Dealing with antimicrobial resistance. Institute of Food Technologists. pp 24

Duffy G, Walsh C, Blair IS, McDowell DA (2006) Survival of antibiotic resistant and antibiotic sensitive strains of E. coli O157 and E. coli O26 in food matrices. Int J Food Microbiol 109:179–186

Fantelli K, Stephan R (2001) Prevalence and characteristics of Shigatoxin-producing Escherichia coli and Listeria monocytogenes strains isolated from minces meat in Switzerland. Int J Food Microbiol 70:63–69

Hajian S, Rahimi E, Mommtaz H (2011) A 3-year study of Escherichia coli O157:H7 in cattle, camel, sheep, goat, chicken and beef minced meat in Proceedings of the International Conference on Food Engineering and Biotechnology (IPCBEE '11). IACSIT Press, Singapoore

Hiko A, Asrat D, Zewde G (2008) Occurrence of *Escherichia coli* O157:H7 in retail raw meat products in Ethiopia. J Infect Dev Ctries 2(5):389–393

Huang L, Juneja VK (2003) Thermal inactivation of Escherichia coli O157: H7 in ground beef supplemented with sodium lactate. J Food Prot 66:664–667

Institute of Food Technologists (IFT) (2003) Expert Report on Emerging Microbiological Food Safety Issues. In: Implications for Control in the 21st Century. S. Lowry/Univ. Ulster/Stone, Institute of Food Technologists, pp 1-32

International Organization for Standardization 17604 (ISO 17604) (2005) Microbiology of Food Animal, Feeding Stuffs-Carcass Sampling for Microbiological Analysis., pp pp 1–pp 12

Kaper JB, Nataro JP, Mobley HLT (2004) Pathogenic Escherichia coli. Nat Rev Microbiol 2:123–140

Khachatourians G (1998) Agricultural Use of Antibiotics and the Evolution and Transfer of Antibiotic Resistant Bacteria. CMAJ 159:1129–1136

Kim H, Samadpour M, Grimm L, Clausen C, Besser T, Baylor M, Kobayashi J, Neill LM, Schoenknecht F, Tarr P (1994) Characteristics of Antibiotic-Resistant Escherichia coli O157:H7 in Washington State, 1984–1991. J Infect Dis 170:1606–1609

Mashood AR, Minga U, Machugun RK (2006) Current Epidemiologic Status of Enterohaemorrhagic Escherichia coli O157:H7 in Africa. Chin Med J 119 (3):217–222

McEvoy JM, Sheridan JJ, McDowell DA (2004) Major pathogens associated with the processing of beef. In: Collins JD (ed) Smulders FJM. Wageningen Academic Publishers, Safety Assurance during Food Processing, pp 57–80

McEwen SA, Fedorka-Cray PJ (2002) Antimicrobial Use and Resistance in Animals. Clin Infect Dis 34(Suppl3):S93–S106

Mead PS, Griffin PM (1998) Escherichia coli O157:H7. Lancet 352(9135):1207–1212

Meng J, Doyle MP (1998) Microbiology of shiga toxin producing *Escherichia coli* in foods. In: Kaper JB, O'Brien AD (eds) *Escherichia coli* O157:H7 and other shiga-toxin producing *E. coli* strains. American Society for Microbiology, Washington, DC, pp pp 92–pp 108

Mølbak K, Mead PS, Griffin PM (2002) Antimicrobial therapy in patients with *Escherichia coli* O157:H7 infection. JAMA 288:1014–1016

Oosterom J (1991) Epidemiological Studies and Proposed Preventive Measures in the Fight Against Human Salmonellosis. Int J Food Microbiol 12:41–51

Pal M (2007) Zoonoses, 2nd edn. Satyam Publishers, Jaipur, India, pp 104–105

Quinn P, Carter M, Markey B, Carter G (2002) Clinical Veterinary Microbiology. Enterobacteriaceae. Wolfe Pub. Spain, In, pp 209–236

Rahimi E, Nayebpour F (2012) Antimicrobial resistance of *Escherichia coli* O157: H7/NM isolated from feces of ruminant animals in Iran. JCAB 6:104–108. doi:10.5897/JCAB11.082

Rangel PM, Marin JM (2009) Antimicrobial resistance in brazilian isolates of Shiga toxin-encoding *Escherichia coli* from cows with mastitis. ARS VETERINARIA, Jaboticabal, SP 25:018–023

Schroeder CM, Zhao C, DebRoy C, Torcolini J, Zhao S, White GD, Wagner DD, McDermott FP, Walker DR, Meng J (2002) Antimicrobial Resistance of Escherichia coli O157:H7 Isolated from Humans, Cattle, Swine, and Food. J Appl Environ Microbiol 68:576–581

Shiomi M, Togawa M, Fujita K, Murata R (1999) Effect of early oral fluroquinones in hemoragic colitis do to *Escherichia coli* O157:H7. Pediatr Int 41(2):228–232

SPSS (2006) Inc SPSS for windows (version 15.0). SPSS Inc., Chicago, Illinois, USA, http://www.hist.umn.edu/~ruggles/hist5011/SPSSBriefGuide150.pdf Accessed 7 July 2014

The Center for Food Security and Public Health (CFSPH) (2009) Enterohemorrhagic Escherichia coli (EHEC) infections. http://www.cfsph. iastate.edu/Factsheets/pdfs/e_coli.pdf Accessed 12 Oct 2013

Vernozy-Rozand C (1999) Verotoxin-producing Escherichia coli (VTEC) and Escherichia coli O157:H7 in medicine and food industry. Ann Biol Clin 57:507–515

Wong CS, Jelacic S, Habeeb RL, Watkins SL, Tarr PI (2000) The risk of hemoragic uramic syndrome after antibiotic treatment of Escherichia coli O157:H7 infections. N Engl J Med 342:1930–1936

3

Estimated the radiation hazard indices and ingestion effective dose in wheat flour samples of Iraq markets

Ali Abid Abojassim[*], Husain Hamad Al-Gazaly and Suha Hade Kadhim

Abstract

In this research, Uranium (^{238}U), Thorium (^{232}Th) and Potassium (^{40}K) specific activity in (Bq/kg) were measured in (12) different types of wheat flours that are available in Iraqi markets. The gamma spectrometry method with a NaI(Tl) detector has been used for radiometric measurements. Also in this study we have calculated the radiation hazard indices (radium equivalent activity and internal hazard index) and Ingestion effective dose in all samples.

It is found that the specific activity in wheat flour samples were varied from (1.086 ± 0.0866) Bq/kg to (12.532 ± 2.026) Bq/kg, for ^{238}U, For ^{232}Th From (0.126 ± 0.066) Bq/kg to (4.298 ± 0.388) Bq/kg and for ^{40}K from (41.842 ± 5.875) Bq/kg to (264.729 ± 3.843) Bq/kg. Also, it is found that the of radium equivalent activity and internal hazard index in wheat flour samples ranged from (3.4031) Bq/kg to (35.1523) Bq/kg and from (0.0091) to (0.1219) respectively. But The range of summation of the Ingestion effective dose were varied from (0.0317) mSv/y to (0.5734) mSv/y. This study prove that the natural radioactivity, radiation hazard indices and Ingestion effective dose were lower than the safe.

Keywords: Wheat flour; Natural radioactivity; Iraq market and gamma spectroscopy

Review

The world is naturally radioactive and approximately 82% of human-absorbed radiation doses, which are out of control, arise from natural sources such as cosmic, terrestrial, and exposure from inhalation or intake radiation sources. In recent years, several international studies have been carried out, which have reported different values regarding the effect of background radiation on human health.

Introduction

Natural radioactivity is caused by the presence of natural occurring radioactive matter (NORM) in the environment. Examples of natural radionuclides include isotopes of potassium (^{40}K), uranium (^{238}U and its decay series), and thorium (^{232}Th and its decay series). In addition to being long-lived (in the order of 10^{10} years), these radionuclides are typically present in air, soil, and water in different amounts and levels of activity. Natural radionuclides are found in terrestrial and aquatic food chains, with subsequent transfer to humans through ingestion of food. As such, international efforts were brought together collaboratively to apply adequate procedures in investigating radionuclides in food (IAEA, International Atomic Energy Agency, Measurements of Radionuclides in Food and Environment 1989), and to set essential guidelines to protect against high levels of internal exposure that may be caused by food consumption (ICRP 1996; UNSCEAR 2000).

Since wheat flour is one of the essential foods that is consumed in Iraqis daily lives, the desire to establish a national baseline of radioactivity exposure from different types of wheat flour samples that available in Iraq markets is very critical. Wheat flour is a powder made from the grinding of wheat used for human consumption. Wheat flour, the "Staff of Life", has been an essential commodity to human existence through the centuries and is currently the most widely consumed staple food. Moreover, numerous studies were conducted worldwide to investigate natural radionuclides in food consumed in different parts of the world (Hosseini et al. 2006; Jibiri & Okusanya 2008; Ababneh et al. 2009; Desimoni et al. 2009). For a systematic treatment, a methodical approach is undertaken that focuses on a wheat flour type of food per study. Because wheat flour is popular among all ages, the current study

* Correspondence: ali.alhameedawi@uokufa.edu.iq
Department of Physics, Kufa University, Faculty of Science, Kufa, Iraq

focuses on investigating the natural radioactive content in all times of food.

Material and methods
Sample collection and preparation
Twelve samples of the most available types of flour were collected from the local markets in Iraq to measure natural activity. The types of samples are listed in Table 1. After collection, each flour sample was kept in a plastic bag and labeled according to its name. All of wheat flour samples were weighed and then dried in an oven at 105°C overnight and reweighed to find the water content. The samples were crushed and were made to pass through a 0.5-mm sieve. Sieved samples were weighed and a mass of 600 g of each sample was placed in a plastic container. The plastic containers were hermetically sealed with adhesive tape for 30 days for secular equilibrium to take place (Nasim et al. 2012).

Measurement system
Natural radioactivity levels were measured using a gamma spectrometer which includes gamma multichannel analyzer equipped with NaI(Tl) detector of (3″ × 3″) crystal dimension as Figure 1. The gamma spectra were analyzed using the ORTEC Maestro-32 data acquisition and analysis system. An energy calibration for this detector is performed with a set of standard gamma ray 37000 Bq active ^{137}Cs, ^{60}Co,^{54}Mn and ^{22}Na sources from USNRC and State License Expert Quantities, "Gamma Source Set", Model RSS- 8. The detector had coaxial closed-facing geometry with the following specifications: The calculated resolution is 7.9% for energy of 661.66 keV of ^{137}Cs standard source. Relative efficiency at 1.33 MeV ^{60}Co was 22% and at 1.274 MeV ^{22}Na was 24%. The lowest limit of detection (LLD) for ^{238}U, ^{232}Th and ^{40}K were 10.86 Bq/kg, 0.569 Bq/kg and 0.0261 Bq/kg respectively. The detector was shielded

Table 1 Types and origin of wheat flour samples in this study

No.	Sample code	Name of Samples	Origin of samples
1	F1	Good sentences	Lebanon
2	F2	Fine semolina	Saudi Arabia
3	F3	Altunsa	Turkey
4	F4	Sirage	Turkey
5	F5	Barrash	Turkey
6	F6	Rehab	IRAQ
7	F7	Sankar	Turkey
8	F8	Super	Turkey
9	F9	Donya	Turkey
10	F10	Suphan	Turkey
11	F11	Farina	Turkey
12	F12	Sayf	Turkey

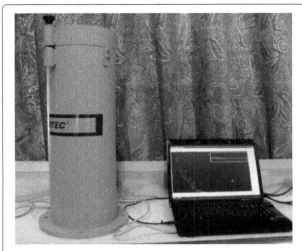

Figure 1 Block diagram of the equipment's set up of NaI(Tl) detector.

by a cylindrical lead shield in order to achieve the lowest background level. An energy calibration for this detector was performed with a set of standard γ-ray 37000 Bq active ^{137}Cs, ^{60}Co,^{54}Mn, and ^{22}Na sources. In this study, the activity concentration of ^{40}K was determined directly from the peak areas at 1460 keV. The activity concentrations of ^{238}U and ^{232}Th were calculated assuming secular equilibrium with their decay products. The gamma transition lines of ^{214}Bi (1765 keV) were used to calculate activity concentration of radioisotope in the ^{238}U-series. The activity concentrations of radioisotope in the ^{232}Th-series were determined using gamma transition lines of ^{208}Tl (2614 keV). The counting time for each sample was at 18000 sec.

Calculation of activity
Since the counting rate is proportional to the amount of the radioactivity in a sample, the Activity Concentration (Ac) which can be determined as a specific activity as the follows (Maduar & Junior 2007):

$$A_C = \frac{C-BG}{\varepsilon \%_c \, MtI_\gamma} \tag{1}$$

Where Ac is the specific activity in (Bq/kg), C is the area under the photo peaks, ε% : Percentage of energy efficiency. I_γ is the percentage of gamma-emission probability of the radionuclide under consideration, t is counting time in (Sec.), M is mass of sample in (kg) and BG is background.

Radium equivalent activity
Radium equivalent activity (Ra_{eq}) is used to assess the hazards associated with materials that contain ^{238}U, ^{232}Th and ^{40}K in Bq/kg (Nasim et al. 2012), which is, determined by assuming that 370 Bq/kg of ^{226}Ra or 260 Bq/kg of ^{232}Th or 4810 Bq/kg of ^{40}K produce the same γ dose rate. The

Ra_{eq} of a sample in (Bq/kg) can be achieved using the following relation (Nasim et al. 2012; Singh et al. 2005; Yu et al. 1992):

$$\mathbf{Ra_{eq}(Bq/kg)} = \mathbf{A_U} + \mathbf{(1.43 \times A_{Th})} + \mathbf{(A_k \times 0.077)} \qquad (2)$$

Internal hazard index

This hazard can be quantified by the internal hazard index (H_{in}) (Nasim et al. 2012; El-Arabi 2007; Quindos et al. 1987). This is given by the following equation:

$$\mathbf{H_{in} = (A_U/185) + (A_{Th}/259) + (A_K/4810)} \qquad (3)$$

The internal hazard index should also be less than one to provide safe levels of radon and its short-lived daughters for the respiratory organs of individuals living in the dwellings.

Ingestion effective dose

The Ingestion effective dose due to the intake of ^{238}U, ^{233}Th and ^{40}K in foods can be evaluated using the following expression: (ICRP 1995; Janet Ayobami 2014).

$$H_{T,r} = \sum_i \left(U_i * C_{i,r} \right) * g_{T,r} \qquad (4)$$

where, i denotes a food group, the coefficients U_i and $C_{i,r}$ denote the consumption rate (kg/y) and activity concentration of the radionuclide r of interest (Bq/kg), respectively, and $g_{T,r}$ is the dose conversion coefficient for ingestion of radionuclide r (Sv/Bq) in tissue T. For adult members of the public, the recommended dose conversion coefficient $g_{T,r}$ for ^{40}K, ^{226}Ra(^{238}U), and ^{232}Th, are 6.2×10^{-9}, 2.8×10^{-7} and 2.2×10^{-7} Sv/Bq respectively (IAEA 1996).

The average consumption rate of wheat flour according to report of ministry of trade in Iraq for adults is 110 Kg/y (Source : The Iraqi Ministry of Trade).

Results and discussion

The specific activity due to ^{238}U, ^{232}Th and ^{40}K in different kinds of wheat flour samples has been measured as shown in Table 2. The specific activity of ^{238}U was found in the range of (1.086 ± 0.0866) Bq/kg to (12.532 ± 2.026)Bq/kg with an average (6.603 ± 3.715) Bq/kg, ^{232}Th from (0.126 ± 0.066)Bq/kg to (4.298 ± 0.388)Bq/kg with an average (1.9465 ± 1.331)Bq/kg and ^{40}K from (41.842 ± 5.875) Bq/kg to (264.729 ± 3.843)Bq/kg with an average (133.097 ± 67.044) Bq/kg.

There is a variation in the specific activity of radionuclides in different wheat flour samples, for example (F1) which is Turkish Farina has lowest ^{238}U concentration, while (F11) which is Lebanese Good sentences has the maximum value, (F8) Turkish Super has the lowest ^{232}Th concentration while the maximum is (F7) also Turkish

Table 2 Specific activity of ^{238}U, ^{232}Th and ^{40}K in wheat flour samples

Sample Code	Specific activity in (Bq/Kg)		
	^{238}U	^{232}Th	^{40}K
F1	1.086 ± 0.0866	3.411 ± 0.322	179.089 ± 3.187
F2	9.991 ± 1.715	3.340 ± 0.356	264.729 ± 3.843
F3	3.391 ± 2.241	0.796 ± 0.504	96.509 ± 2.446
F4	5.102 ± 1.861	2.462 ± 0.475	120.555 ± 5.5134
F5	2.243 ± 2.303	1.646 ± 0.394	47.805 ± 5.025
F6	6.599 ± 1.852	1.375 ± 0.655	100.892 ± 6.289
F7	11.078 ± 2.848	4.298 ± 0.388	79.767 ± 6.499
F8	BLD	0.126 ± 0.066	41.842 ± 5.875
F9	6.048 ± 1.526	1.561 ± 0.664	109.061 ± 6.643
F10	6.196 ± 3.127	1.652 ± 0.684	191.549 ± 7.006
F11	12.532 ± 2.026	2.685 ± 0.573	175.257 ± 6.510
F12	6.370 ± 2.307	BLD	190.104 ± 7.876
Average ± S.D	6.603 ± 3.715	1.9465 ± 1.331	133.097 ± 67.044

Sankar , and the lowest ^{40}K concentration is (F8) which is Turkish Super and the maximum is (F2) Saudi Arabia Fine semolina. The results obtained show that the specific activity of ^{238}U, ^{232}Th and ^{40}K in all wheat flour samples appeared lower than recommended limit of UNSCEAR (United Nations Scientific Committee on Effects of Atomic Radiation (UNSCEAR) 2008).

The radiation hazard indices (radium equivalent activity and internal hazard indices) were calculated for all samples in this study as shown in Table 3).The radium equivalent activity internal hazard indices were varied from (3.4031) to (35.1523) with an average (19.6347 ± 9.1680) and from

Table 3 Radium equivalent activity and internal hazard index in wheat flour samples

Sample code	Ra_{eq} (Bq/kg)	H_{in}
F1	28.3442	0.1027
F2	35.1523	0.1219
F3	11.9621	0.0414
F4	17.9069	0.0621
F5	8.2789	0.0284
F6	16.3357	0.0619
F7	23.3670	0.0931
F8	3.4031	0.0091
F9	16.6801	0.0614
F10	23.3093	0.0797
F11	29.8681	0.1145
F12	21.0081	0.07395
Average ± S.D	19.6347 ± 9.1680	0.0708 ± 0.0341

Table 4 Ingestion effective dose for adult in wheat flour samples

Sample Code	Ingestion effective dose (mSv/y)			
	^{238}U	^{232}Th	^{40}K	Sum
F1	0.0334	0.0863	0.1221	0.2419
F2	0.3077	0.0845	0.1805	0.5728
F3	0.1044	0.0201	0.0658	0.1904
F4	0.1571	0.0623	0.0822	0.3016
F5	0.0691	0.0416	0.0326	0.1433
F6	0.2032	0.0348	0.0688	0.3068
F7	0.3412	0.1087	0.0544	0.5043
F8	BLD	0.0032	0.0285	0.0317
F9	0.1863	0.0395	0.0743	0.3001
F10	0.1908	0.0418	0.1306	0.3633
F11	0.3859	0.0679	0.1195	0.5734
F12	0.1962	BLD	0.1297	0.3258
Average ± S.D	0.1978 ± 0.1066	0.0537 ± 0.0317	0.0908 ± 0.0457	0.3213 ± 0.1657

(0.0091) to (0.1219) with an average (0.0708 ± 0.0341) respectively.

The values of all the radiation hazard indices in this study (radium equivalent activity and internal hazard indices are lowest value in sample (F8) Turkish Super and the highest value in sample (F2) Saudi Arabia Fine semolina. This indicates that the internal hazard index in wheat flour samples were lower than the permissible limits of 1 recommended by UNSCEAR (UNSCEAR 2000), while the radium equivalent activity also were lower than the maximum permissible level of 370 Bq/kg recommended by UNSCEAR (UNSCEAR 2000).

Table 4 shows the results of the Ingestion effective dose in (mSv/y) for adult due to specific activity of ^{238}U, ^{232}Th and ^{40}K in wheat flour samples which it is calculated using Eq. (4). The range of summation of the Ingestion effective dose were varied from (0.0317) mSv/y (at sample F8) to (0.5734) mSv/y (at sample F11) with an average (0.3213 ± 0.1657) mSv/y, but Figure 2 shows the compare between average of the Ingestion effective dose for ^{238}U, ^{232}Th and ^{40}K in wheat flour samples which obtain the average of Ingestion effective dose due to ^{238}U was higher than due to ^{232}Th and ^{40}K because of the increased the dose conversion coefficient for ingestion of radionuclide. This indicates that the Ingestion effective dose in all wheat flour samples were lower than the permissible limits of 1 mSv/y recommended by ICRP (ICRP 1996).

Conclusion

The present study has presented the specific activity of radionuclides ^{238}U, ^{232}Th and ^{40}K using gamma ray spectroscope in different type of wheat flour that are regularly consumed by adults age in Iraq. Specific activity concentrations of these radionuclides in samples were lower than as reported by UNSCEAR. Also the radium equivalent activity and internal hazard indices values obtained when compared with the world permissible values were found to be below the standards limit which due to be radiologically hazard safe. The high value of summation of Ingestion effective was less than 1 mSv/y, the limit recommended for the public (ICRP 1996), hence wheat flour samples in Iraq markets products are safe to consumers.

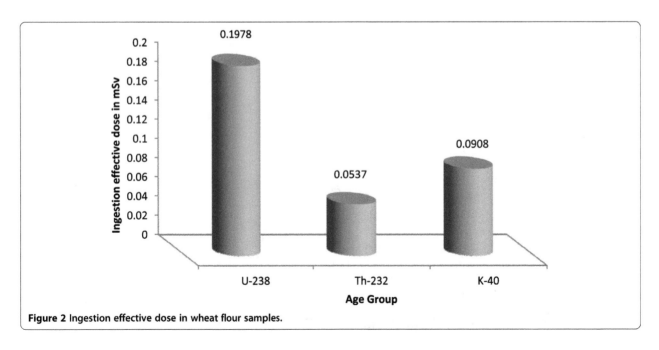

Figure 2 Ingestion effective dose in wheat flour samples.

Competing interests
The authors declare that they have no competing interests.

Authors' contributions
AAA and HHA-G carried out the Nuclear radiation studies, participated in the sequence alignment and drafted the manuscript. SHK collected and arranged wheat flour samples, also contributed to the collection of references of scientific. All authors read and approved the final manuscript.

Acknowledgements
I would like to knowledge all those contributed in declaring this issue. Special thanks to the staff of the department of physics at Kufa University.

References
Ababneh ZQ, Alyassin AM, Aljarrah KM, Ababneh AM (2009) Measurement of natural and artificial radioactivity in powdered milk consumed in Jordan and estimates of the corresponding annual effective dose. Radiat Prot Dosimetry 138:278–283

Desimoni J, Sives F, Errico L, Mastrantonio G, Taylor MA (2009) Activity levels of gamma-emitters in Argentinean cow milk. J Food Compos Anal 22:250–253

El-Arabi A (2007) ^{226}Ra, ^{232}Th and ^{40}K concentrations in igneous rocks from eastern desert, Egypt and its radiological implication. Radiation Measurement 42:94–100

Hosseini T, Fathivand AA, Abbasisiar, Karimi M, Barati H (2006) Assessment of annual effective dose from U-238 and Ra-226 due to consumption of foodstuffs by inhabitants of Tehran city, Iran. Radiat Prot Dosim 121:330–332

IAEA (1996) International Atomic Energy Agency, International Basic Safety Standard for Protection against Ionizing Radiation and for the Safety of Radiation Sources. Series No. 115, International Atomic Energy Agency (IAEA), Vienna

IAEA, International Atomic Energy Agency (1989) Measurement of Radionuclides in Food and the Environment. Technical Report Series No. 295, Vienna

ICRP (1995) Age-dependent Doses to the Members of the Public from Intake of Radionuclides - Part 5 Compilation of Ingestion and Inhalation Coefficients. ICRP Publication 72. Annex ICRP 26 (1)

ICRP (1996) International Commission on Radiological Protection, Age-Dependent Doses to Members of the Public from Intake of Radionuclides: Part 5 Compilations of In-gestion and Inhalation Dose Coefficients (ICRP Publica-tion 72)". Pergamon Press, Oxford

Janet Ayobami A (2014) Estimation of Annual Effective Dose Due to Ingestion of Natural Radionuclides in Cattle in Tin Mining Area of Jos Plateau, Nigeria. Nat Sci 6:255–261

Jibiri NN, Okusanya AA (2008) Radionuclide contents in food products from domestic and imported sources in Nigeria. J Radiol Prot 28:405–413

Maduar M, Junior P (2007) Gamma Spectrometry In the Determination of Radionuclides Comprised In Radioactivity Series, International Nuclear Atlantic Conference-INC. Santos SP, Brazil

Nasim A, Sabiha J, Tufail M (2012) Enhancement of natural radioactivity in fertilized soil of Faisalabad, Pakistan. Environ SciPollut Res 19:3327–3338

Quindos L, Fernandez P, Soto J (1987) Building materials as source of exposure in houses. In: Seifert B, Esdorn H (eds) Indoor Air 87. Institute of Water, Soil and Air Hygiene, Berlin, p 365

Singh S, Rani A, Mahajan R (2005) ^{226}Ra, ^{232}Th and ^{40}K analysis in soil samples from some areas of Punjab and Himachal Pradesh, India using gamma ray spectrometry. Radiation Measurement 39:431–439

United Nations Scientific Committee on Effects of Atomic Radiation (UNSCEAR) (2008) Report to the General Assembly. Sources and Effects of Ionizing Radiation, New York

UNSCEAR (2000) Sources and effects of ionizing radiation. United Nations Scientific Committee on the Effects of Atomic Radiation Effects of Atomic Radiation. Report to the General Assembly with annexes. United Nations, New York

Yu K, Guan Z, Stoks M, Young E (1992) The assessment of natural radiation dose committed to the Hong Kong people. J Environ Radioact 17:31–48

4

Dietary exposure to total and inorganic arsenic in the United States, 2006–2008

Elizabeth A Jara and Carl K Winter[*]

Abstract

Background: Consumers are frequently exposed to arsenic in foods and considerable public and scientific concern exists regarding the potential health risks from dietary arsenic. Arsenic exists in both organic and inorganic forms and health effects are primarily attributed to inorganic forms. The most common analytical methods used to detect arsenic measure total arsenic, which includes both organic and inorganic forms. It is therefore necessary to make assumptions concerning the amounts of total arsenic found in food samples that represent inorganic arsenic. This work presents a new assessment of US dietary exposure to arsenic using data available from the FDA Total Diet Study from 2006-2008 and a series of scenarios developed to estimate inorganic arsenic levels.

Results: Total arsenic exposures were estimated for 16 population subgroups and ranged from 1.4×10^{-1} to 4.5×10^{-1} μg/kg/day. The population subgroup with the highest exposure to total arsenic was 2 year-old children. The major food group contributors to total arsenic exposure for the general US population were marine sources, which accounted for 69 percent of the total arsenic exposure, and grains, legumes and seeds, which accounted for 20 percent. The highest inorganic arsenic exposures occurred for 2 year-old children and ranged from to 1.1×10^{-1} μg/kg/day to 2.4×10^{-1} μg/kg/day. Inorganic arsenic exposures for the 2 year-olds were 3.3 to 4.8 times higher than inorganic arsenic exposures for the general population. Under Scenario 5, which assumed that 70 percent of total arsenic from terrestrial sources and 10 percent of total arsenic from marine sources existed as inorganic arsenic, the most important food group contributors to inorganic arsenic for 2 years-olds were grains, legumes and seeds (50 percent), beverages (14 percent), marine sources (13 percent) and snacks and breakfast cereals (12 percent).

Conclusions: The exposures estimates obtained in this work are below the EPA's established RfD of 0.3 μg/kg/day for inorganic arsenic and below EFSA's health concern level of 0.3 to 8.0 μg/kg/day. To improve the accuracy of future arsenic risk assessments, studies should incorporate specific analytical data measuring inorganic arsenic from foods contributing the most to inorganic arsenic exposure.

Keywords: Dietary exposure; Arsenic; Speciation; Food safety

Background

Arsenic is an element distributed in soil, water and air. It is frequently detected in drinking water and food and occurs in the items from natural and anthropogenic sources (EPA 2013; FDA 2013a). Arsenic exists in both organic and inorganic forms and a wide variety of arsenical species have been identified. Inorganic arsenic forms as arsenite (As^{3+}) and arsenate (As^{5+}) are the most harmful forms in health terms and public concern (ATSDR 2005).

The US Environment Protection Agency (EPA) and the International Agency for Research on Cancer (IARC) consider arsenic as a known human carcinogen (EPA 2012a; IARC 2012). In addition to carcinogenicity, arsenic also exerts a variety of acute (short-term) and chronic (long-term) toxicological effects. Acute effects from arsenic exposure have resulted principally from occupational exposure and include gastrointestinal effects such as nausea, diarrhea and abdominal pain (ATSDR 2005). Chronic exposure to arsenic, in addition to bladder, lung and liver cancer, has also been linked to skin toxicity, hyperpigmentation, anemia, peripheral neuropathy and neurocognitive problems (WHO 2001).

* Correspondence: ckwinter@ucdavis.edu
Department of Food Science and Technology, University of California, Davis, CA, USA

Human exposure to arsenic is a complex issue because it is closely related to environmental pollution, occupation, lifestyle and dietary patterns of the exposed population. While drinking water has traditionally been considered as a principal contributor of consumer exposure to arsenic, recent studies have shown that food represents an even greater source of consumer exposure to arsenic (Georgopolous et al. 2008; Xue et al. 2010). Fruit juices and rice have been identified as key contributors to arsenic exposure in the population and the presence of arsenic in these foods has been the subject of considerable media coverage (Tavernise 2013a, b). Specific attention has been focused upon exposure to children below five years of age due to their significant consumption of these foods.

EPA has set an arsenic safety standard for drinking water at 10 parts per billion (ppb) while the US Food and Drug Administration (FDA) has set an action level of 10 ppb arsenic in apple juice (EPA 2012b; FDA 2013b). Other foods, however, are not subject to arsenic action levels and include rice and rice cereals.

According to EPA, the Reference Dose (RfD) for inorganic arsenic is 0.3 micrograms per kilogram body weight per day (µg/kg/day) (EPA 2012a). The RfD is an estimate of a daily oral exposure to the human population that is likely to be without appreciable risk of detrimental noncancer effects during a lifetime. The RfD for inorganic arsenic is based on hyperpigmentation, keratosis, and possible vascular complications in humans including sensitive groups (EPA 2012a).

The Joint Food and Agricultural Organization/World Health Organization (FAO/WHO) Expert Committee on Food Additives (JECFA) in 2011 withdrew the provisional weekly intake of 15 µg/kg/day for inorganic arsenic exposure after concluding that more recent studies have shown that a range of adverse effects have been reported at exposures lower than those used when establishing the provisional weekly intake. Based on epidemiological studies, JEFCA considered inorganic arsenic exposures between 2.0 to 7.0 µg/kg/day as levels of concern for potential cancers of the skin, lung and bladder (FAO/WHO 2011).

Studies conducted by the FDA have resulted in dietary estimates of 1.3 to 12.5 µg inorganic arsenic per day (Tao and Bolger 1999). These exposures represent 0.026 to 0.25 µg/kg/day for a 50 kg person. A study by the EPA provided a mean inorganic arsenic exposure of 0.05 µg/kg/day for the overall population and ranges of 0.08 to 0.23 µg/kg/day for children under 5 years of age (Xue et al. 2010).

Estimates of consumer exposure to inorganic arsenic from foods have also been made by the European Food Safety Authority (EFSA). In 2009, inorganic arsenic exposure estimates ranged from 0.13 to 0.56 µg/kg/day for average consumers and from 0.37 to 1.22 µg/kg/day for 95th percentile consumers; the dietary exposure to inorganic arsenic for children under three years of age was estimated to be from 2 to 3 times that of adults (EFSA 2009). In 2014, using different methods to consider arsenic speciation, EFSA reported that mean inorganic arsenic dietary exposure estimates ranged from 0.20 to 1.37 µg/kg/day for infants, toddlers, and other children and from 0.09 to 0.38 µg/kg/day for the adult population. The 95th percentile dietary exposure estimate ranged from 0.36 to 2.09 µg/kg/day for infants, toddlers, and other children. For adults, the 95th percentile dietary exposure ranged from 0.14 to 0.64 µg/kg/day (EFSA 2014).

The toxicity of arsenic to humans is directly related to the form (inorganic vs. organic) that the arsenic exists in, with inorganic forms of arsenic presenting the greatest concern. Unfortunately, due to the high cost and the lack of validated analytical methods for arsenic speciation, most arsenic exposure assessments have relied upon measurements of total arsenic from food items that do not differentiate between organic and inorganic arsenic.

In some cases, differences between total and inorganic arsenic measured from foods can be significant. As an example, foods of marine origin (finfish, shellfish, algae) represent significant sources of dietary arsenic exposure although US studies have indicated that less than 10 percent of the total arsenic in finfish and approximately 30 percent of the total arsenic in shellfish from uncontaminated waters represented inorganic arsenic (Valette-Silver et al. 1999; Lorenzana et al. 2009). Worldwide literature shows similar results with 7.3 percent of total arsenic for finfish and 25 percent of total arsenic for shellfish present in the inorganic form (Mohri et al. 1990; Suñer et al. 1999 2002; Muñoz et al. 2000).

Accurate human exposure assessments for dietary inorganic arsenic require that speciation differences among inorganic and organic forms of arsenic be appropriately considered which presents a significant challenge due to the lack of speciation data. The 1999 FDA inorganic arsenic exposure assessment was performed using the best arsenic contamination data available at the time. In the absence of specific inorganic arsenic data, it was assumed that 10 percent of the total arsenic from all marine sources was considered to be present as inorganic arsenic while 100 percent of the total arsenic from all other sources was in the inorganic arsenic form (Tao and Bolger 1999).

The 2009 EFSA report considered various scenarios relating to the proportion of total arsenic representing inorganic arsenic based upon reasonable assumptions concerning arsenic speciation of various foods obtained from EFSA monitoring and in the scientific literature. For fish and other seafood, levels of inorganic arsenic were obtained using either food contamination data, fixed upper bound estimates, or fixed lower bound estimates. For all other food categories scenarios were developed to reflect either 100 percent, 70 percent, or 50 percent of the total arsenic representing inorganic arsenic. The cross combination of these assumptions produced nine scenarios characterizing inorganic arsenic in marine and terrestrial food sources (EFSA 2009). The recent 2014 EFSA exposure assessment, in contrast, used only data specifically measuring inorganic arsenic levels from fish and other seafood while assuming 70 percent of the total arsenic found in all other foods represented inorganic arsenic (EFSA 2014).

Using a traditional probabilistic exposure assessment method combined with urinary biomarker data examining various forms of inorganic arsenic, EPA researchers concluded that approximately 10 percent of arsenic exposure from foods was in the toxic inorganic form (Xue et al. 2010).

FDA recently published data concerning levels of inorganic arsenic detected from fruit juices and from rice products, including rice, rice cereals, snacks, pasta, grain-based bars and bakery mixtures (FDA 2013c). Unfortunately, arsenic has also been found in numerous other foods using analytical techniques incapable of differentiating organic and inorganic arsenic. As a result, dietary human inorganic arsenic exposure assessments still require assumptions to be made about the relative prevalence of inorganic arsenic compared with total arsenic. The most recent FDA assessment of total dietary arsenic exposure was published in 1999 and reflected arsenic measurements from food taken between 1991 and 1996 (Tao and Bolger 1999).

This paper provides an assessment of US dietary exposure to inorganic arsenic by incorporating the most recently available data (2006–2008) obtained from the FDA's Total Diet Study (TDS). A series of scenarios have been developed to estimate the percentage of total arsenic detected that represents inorganic arsenic and to identify key drivers of inorganic arsenic exposure. Exposure estimates are compared with those in the published literature and with human health criteria to determine the significance of such exposures.

Methods

Dietary intake of arsenic in the US population was estimated using results from the FDA Total Diet Study (TDS) Market Baskets 2006–1 through 2008–4. Food consumption estimates for foods analyzed in the TDS were provided by FDA and were derived from the 1994–1996 Continuing Survey of Food Intakes for Individuals (CSFII) developed by the US Department of Agriculture. Each year, FDA inspectors collect four Market Baskets from retail outlets (each from a different geographical location) containing approximately 280 different food items. The food items are prepared for consumption and then analyzed for pesticide residues, industrial chemicals,

radionuclides, and toxic and nutrient elements, including arsenic (FDA 2008).

Arsenic levels detected in the TDS were reported as the total arsenic concentration in each food item (FDA 2010). Dietary exposures to arsenic were calculated by multiplying median food consumption levels by the arsenic concentrations means and divided by body weights (EPA 2011) for each population subgroup. The arsenic concentration means reported from the TDS assumed that no arsenic was present at levels below the limit of detection. Sixteen population subgroups were considered, including all males, all females, infants 6–11 months, children 2 years, children 6 years, children 6–10 years, males (14–16 years, 25–30 years, 40–45 years, 60–65 years, 70 years) and females (14–16 years, 25–30 years, 40–45 years, 60–65 years, 70 years). Total exposure to arsenic for each population subgroup was calculated by combining the arsenic contributions from each food item.

To identify the key food contributors to dietary arsenic exposure, food items were classified into core groups generally consistent with the FDA food list core groups but with some modifications (Pennington 1992; Egan et al. 2007).

Using urinary biomarkers of inorganic arsenic exposure combined with probabilistic arsenic exposure estimates derived from food consumption databases and TDS sampling of foods for total arsenic, it has been estimated that approximately 10 percent of total arsenic in foods is present in the inorganic form (Xue et al. 2010). This approach, while elegant, provides only crude estimates as to the arsenic speciation among specific food groups.

Historically, conservative approaches have been used to assess dietary exposure to inorganic arsenic in food and the majority of them have assumed that 100 percent of terrestrial arsenic is consumed as inorganic arsenic. This assumption may overestimate dietary exposure to inorganic arsenic. On the other hand, some exposure estimates assume all arsenic detected from marine sources exists as organic arsenic, so such approaches could underestimate dietary exposure to inorganic arsenic (Yost et al. 1998; Tao and Bolger 1999).

This work presents additional options to assess arsenic inorganic exposure in food based on the classification of

Table 1 Scenario assumptions: percentage of total arsenic occurring as inorganic arsenic from marine and terrestrial sources

Inorganic arsenic	100% terrestrial	70% terrestrial	50% terrestrial
0% marine sources	Scenario 1	Scenario 4	Scenario 7
10% marine sources	Scenario 2	Scenario 5	Scenario 8
25% marine sources	Scenario 3	Scenario 6	Scenario 9

Table 2 Total arsenic exposure for 16 population subgroups (µg/kg/day)

Core Food groups	A. Milk and cheese	B. Meat and poultry	C. Grains, legumes and seeds	D. Fruits and vegetables	E. Desserts	F. Snacks and breakfast cereals	G. Condiments, sugars and sweeteners	H. Beverages	I. Marine sources	Total exposure
Total US F	1.7E-04	5.5E-03	3.5E-02	3.3E-03	1.3E-03	4.4E-03	6.5E-06	3.5E-03	1.2E-01	1.7E-01
Total US M	1.5E-04	4.7E-03	3.1E-02	2.8E-03	1.1E-03	3.8E-03	5.6E-06	3.0E-03	1.0E-01	1.5E-01
6-11 mo.	1.7E-04	3.6E-03	3.9E-02	2.6E-03	3.8E-03	4.5E-02	7.6E-06	5.3E-02	5.2E-02	2.0E-01
M/F 2 yr	7.1E-04	1.4E-02	1.3E-01	8.5E-03	3.4E-03	3.1E-02	1.2E-05	3.5E-02	2.3E-01	4.5E-01
M/F 6 yr	5.0E-04	1.1E-02	7.8E-02	6.3E-03	5.0E-03	3.0E-02	8.6E-06	9.5E-03	1.7E-01	3.1E-01
M/F 10 yr	3.7E-04	6.5E-03	4.8E-02	3.1E-03	3.5E-03	2.3E-02	9.4E-07	2.0E-03	1.6E-01	2.4E-01
F 14–16 yr	2.0E-04	5.3E-03	3.3E-02	2.2E-03	1.3E-03	5.8E-03	3.0E-06	2.3E-03	1.1E-01	1.6E-01
M 14–16 yr	3.0E-04	6.0E-03	4.1E-02	2.5E-03	2.1E-03	9.5E-03	1.1E-05	1.9E-03	8.5E-02	1.5E-01
F 25–30 yr	1.7E-04	5.0E-03	4.1E-02	3.5E-03	1.2E-03	3.2E-03	3.5E-06	3.5E-03	1.1E-01	1.7E-01
M 25–30 yr	2.3E-04	6.9E-03	3.9E-02	2.4E-03	1.0E-03	2.3E-03	3.9E-06	2.4E-03	9.5E-02	1.5E-01
F 40–45 yr	1.0E-04	4.6E-03	3.1E-02	3.4E-03	1.1E-03	2.7E-03	9.2E-06	3.2E-03	9.7E-02	1.4E-01
M 40–45 yr	1.5E-04	6.0E-03	3.2E-02	2.7E-03	1.2E-03	2.8E-03	1.8E-06	3.2E-03	8.8E-02	1.4E-01
F 60–65 yr	1.0E-04	3.2E-03	2.0E-02	3.5E-03	8.5E-04	3.1E-03	1.5E-05	3.1E-03	1.5E-01	1.9E-01
M 60–65 yr	1.2E-04	4.3E-03	2.7E-02	3.5E-03	1.1E-03	2.8E-03	5.7E-06	3.1E-03	1.3E-01	1.7E-01
F 70 yr	7.8E-05	3.7E-03	1.9E-02	4.1E-03	1.0E-03	3.4E-03	7.6E-06	2.2E-03	1.2E-01	1.5E-01
M 70 yr	9.2E-05	4.1E-03	1.7E-02	3.5E-03	9.6E-04	3.1E-03	1.5E-05	2.7E-03	1.3E-01	1.6E-01

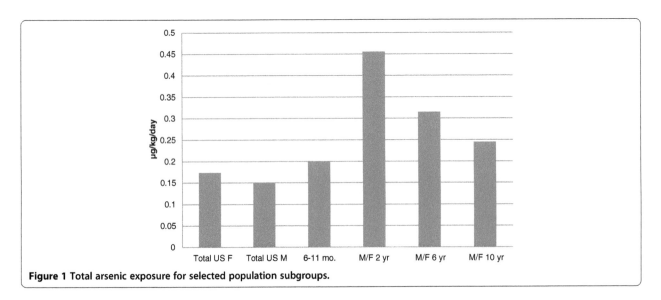

Figure 1 Total arsenic exposure for selected population subgroups.

food in marine and terrestrial sources. We postulate that marine sources contribute to inorganic arsenic exposure and could represent as much as 25 percent of the total arsenic found from marine sources (Muñoz et. al. 2000; Súñer et al. 2002; De Gieter et al. 2002; Lorenzana et al. 2009). Large variability is also seen in the arsenic speciation from food of terrestrial sources. In this study, we estimate that at least 50 percent of total arsenic from terrestrial sources exists as inorganic arsenic (Carbonell-Barrachina et al. 2012; Somella et al. 2013).

In this study, a series of nine scenarios was developed with each scenario making different assumptions of the percentages of total arsenic from marine and terrestrial sources present as inorganic arsenic. The assumptions behind each scenario are presented in Table 1. Scenario 3

presents the worst-case exposure scenario while Scenario 7 presents considers conditions that would result in the lowest exposures. Intermediate exposures are expected from Scenario 5.

Results and discussion
Total arsenic
The total arsenic exposures for the 16 age/sex groups are reported in Table 2. The total exposure estimates ranged from 1.4×10^{-1} to 4.5×10^{-1} µg/kg/day. US female exposure to total arsenic was estimated to be 1.7×10^{-1} µg/kg/day while US male exposure to total arsenic was estimated to be 1.5×10^{-1} µg/kg/day.

The population subgroup with the greatest exposure to total arsenic was 2 year-old children, who were exposed to

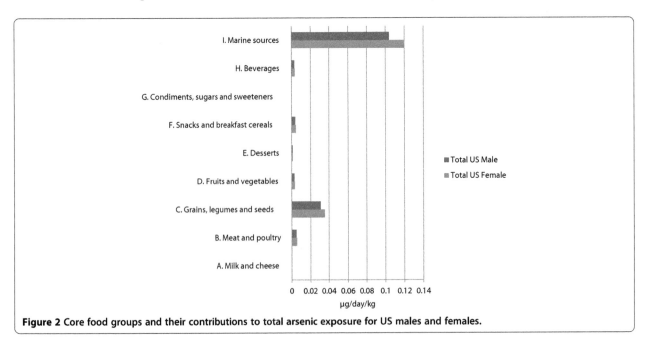

Figure 2 Core food groups and their contributions to total arsenic exposure for US males and females.

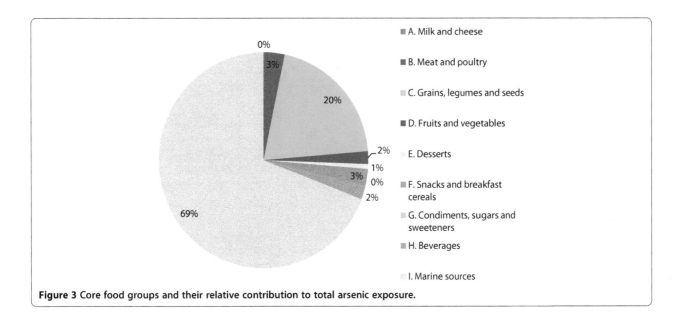

Figure 3 Core food groups and their relative contribution to total arsenic exposure.

4.5×10^{-1} µg/kg/day. Other population subgroups with exposures above at least 2.0×10^{-1} µg/kg/day included 6–11 month infants (2.0×10^{-1} µg/kg/day), 6 year-old children (3.1×10^{-1} µg/kg/day), and 10 year-old children (2.4×10^{-1} µg/kg/day). The lowest total arsenic exposures occurred in the 40–45 year old male and female groups at 1.4×10^{-1} µg/kg/day.

A graph of total arsenic exposure for US males, US females, and the most exposed population subgroups is provided in Figure 1.

The contributions of all of the core food groups to total arsenic exposure in the US population are shown in Figure 2. The top core food group contributing to total arsenic exposure was marine sources, which was responsible for exposures of 1.2×10^{-1} µg/kg/day for females and 1.0×10^{-1} µg/kg/day for males. Grains, legumes and seeds contributed 3.5×10^{-2} µg/kg/day for females and

3.1×10^{-2} µg/kg/day for males. Meat and poultry contributed 5.5×10^{-3} µg/kg/day (female) and 4.7×10^{-3} µg/kg/day (male) while snacks and breakfast cereals contributed 4.4×10^{-3} µg/kg/day and 3.8×10^{-3} µg/kg/day for females and males, respectively.

Figure 3 shows the percentage contribution of the nine core food groups to the total arsenic exposure in the general US population. No significant gender differences were observed. Marine sources were responsible for 69 percent of the total arsenic exposure, followed by grains, legumes and seeds (20 percent). Another three percent of the total arsenic exposure was contributed by both meat and poultry and snacks and breakfast cereals. These top four core food groups were responsible for 95 percent of the total arsenic exposure.

Rice has been identified as a major contributor to dietary arsenic exposure and steps were taken to determine

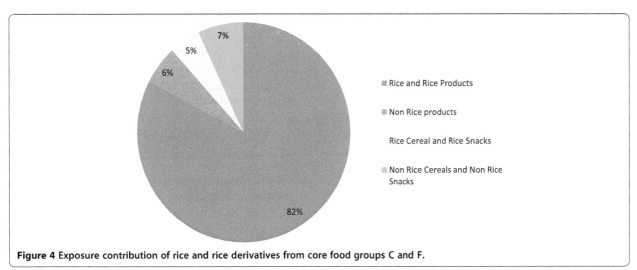

Figure 4 Exposure contribution of rice and rice derivatives from core food groups C and F.

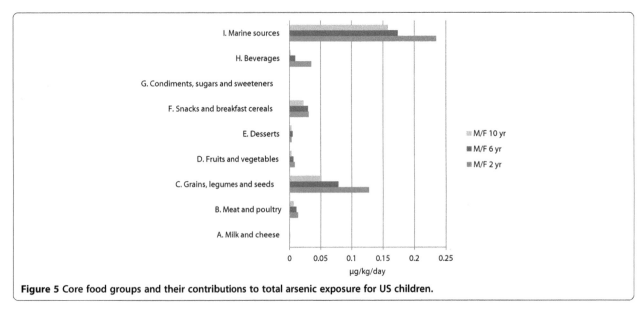

Figure 5 Core food groups and their contributions to total arsenic exposure for US children.

the specific contributions of rice and rice-based foods to total arsenic exposure. Consumption of arsenic from rice and rice-based foods was evident in the grains, legumes, and seeds (rice and rice products) and the snacks and breakfast cereals (rice cereals and rice snacks) core food groups that contributed 23 percent of the total arsenic exposure. To determine the extent by which rice and rice-based foods were responsible for arsenic exposures in these two core food groups, those groups were organized into subcategories including rice and rice products, non-rice products, rice cereals and rice snacks, and non-rice cereals and non-rice snacks. Foods containing rice contributed to 87 percent of the total arsenic exposure from those two core food groups (Figure 4), and the overall contribution of rice and rice-based foods to total arsenic exposure was 20 percent.

Among the three population subgroups receiving the highest total arsenic exposures (2 year-olds, 6 year-olds, 10 year-olds), the relative contributions of the core food groups were similar to those of the total population. Highest exposures (Figure 5) for 6 year-olds and 10 year-olds were from marine sources, followed by grains, legumes and seeds, snacks and breakfast cereals, and beverages. For 2 year-olds, highest total arsenic exposures were also from marine sources and from grains, legumes and seeds, but exposures to total arsenic from consumption of beverages were greater than exposure to total arsenic from snacks and breakfast cereals.

Inorganic arsenic

Inorganic arsenic exposure was estimated for the 16 population subgroups using nine different scenarios. Each scenario considered different assumptions regarding the fraction of total arsenic that was present as inorganic arsenic. Results of the nine scenarios are shown in Table 3.

The highest inorganic arsenic exposure (2.4×10^{-1} µg/kg/day) was observed for the 2 year-old population subgroup under Scenario 3, which assumed that 100 percent of the total arsenic from terrestrial sources and 25 percent of the total arsenic from marine sources was present as inorganic arsenic. Inorganic arsenic exposure for 2 year-olds drops to 1.8×10^{-1} µg/kg/day under Scenario 5 (70 percent terrestrial, 10 percent marine) and to 1.1×10^{-1} µg/kg/day under Scenario 7 (50 percent terrestrial, 0 percent marine). Estimated exposures to inorganic arsenic for the general population were much lower, with male and female exposures ranging from 2.3 to 2.7×10^{-2} µg/kg/day under Scenario 7, 4.3 to 5.0×10^{-2} µg/kg/day under Scenario 5, and 7.2 to 8.4×10^{-2} µg/kg/day under Scenario 3.

Inorganic arsenic exposure estimates for total male, total female, 6–11 month, 2 year-old, 6 year-old, and 10 year-old population subgroups using Scenarios 3, 5, and 7 are presented in Figure 6. Inorganic arsenic exposures for the 2 year-olds ranged from 3.3 to 4.8 times higher than the general population while exposures to the 6 year-olds were 2.7 to 3.0 times higher than the general population and exposures to 10 year-olds were 1.5 to 1.9 times higher than the general population.

The influence of assumptions concerning the percentage of total arsenic present as inorganic arsenic is considered for the 6–11 month, 2 year-old, and 6 year-old population subgroups in Table 4. Under Scenario 3, which considers all total arsenic from terrestrial sources to represent inorganic arsenic and 25 percent of all total arsenic from marine sources to represent inorganic arsenic, inorganic arsenic represents 61 percent of the total arsenic exposure. For Scenario 5 (70 percent terrestrial, 10 percent marine), inorganic arsenic represents 39 percent of the total arsenic exposure while for Scenario

Table 3 Inorganic arsenic exposure for population subgroups based on speciation assumptions (µg/kg/day)

Age/Sex groups	Scenario 1 100% T 0% MS	Scenario 2 100% T 10% MS	Scenario 3 100% T 25% MS	Scenario 4 70% T 0% MS	Scenario 5 70% T 10% MS	Scenario 6 70% T 25% MS	Scenario 7 50% T 0% MS	Scenario 8 50% T 10% MS	Scenario 9 50% T 25% MS
Total US F	5.4E-02	6.6E-02	8.4E-02	3.8E-02	5.0E-02	6.8E-02	2.7E-02	3.9E-02	5.7E-02
Total US M	4.6E-02	5.7E-02	7.2E-02	3.3E-02	4.3E-02	5.9E-02	2.3E-02	3.4E-02	4.9E-02
6-11 mo.	1.5E-01	1.5E-01	1.6E-01	1.0E-01	1.1E-01	1.2E-01	7.4E-02	7.9E-02	8.7E-02
M/F 2 yr	2.2E-01	2.4E-01	2.8E-01	1.5E-01	1.8E-01	2.1E-01	1.1E-01	1.3E-01	1.7E-01
M/F 6 yr	1.4E-01	1.6E-01	1.8E-01	9.8E-02	1.2E-01	1.4E-01	7.0E-02	8.8E-02	1.1E-01
M/F 10 yr	8.7E-02	1.0E-01	1.3E-01	6.1E-02	7.6E-02	1.0E-01	4.3E-02	5.9E-02	8.3E-02
F 14–16 yr	5.0E-02	6.1E-02	7.8E-02	3.5E-02	4.6E-02	6.3E-02	2.5E-02	3.6E-02	5.3E-02
M 14–16 yr	6.3E-02	7.2E-02	8.4E-02	4.4E-02	5.3E-02	6.5E-02	3.2E-02	4.0E-02	5.3E-02
F 25–30 yr	5.7E-02	6.8E-02	8.5E-02	4.0E-02	5.1E-02	6.7E-02	2.9E-02	4.0E-02	5.6E-02
M 25–30 yr	5.4E-02	6.4E-02	7.8E-02	3.8E-02	4.8E-02	6.2E-02	2.7E-02	3.7E-02	5.1E-02
F 40–45 yr	4.6E-02	5.6E-02	7.1E-02	3.3E-02	4.2E-02	5.7E-02	2.3E-02	3.3E-02	4.7E-02
M 40–45 yr	4.8E-02	5.7E-02	7.0E-02	3.4E-02	4.3E-02	5.6E-02	2.4E-02	3.3E-02	4.6E-02
F 60–65 yr	3.4E-02	4.9E-02	7.2E-02	2.4E-02	3.9E-02	6.2E-02	1.7E-02	3.2E-02	5.5E-02
M 60–65 yr	4.2E-02	5.5E-02	7.4E-02	2.9E-02	4.2E-02	6.1E-02	2.1E-02	3.4E-02	5.3E-02
F 70 yr	3.3E-02	4.5E-02	6.3E-02	2.3E-02	3.5E-02	5.3E-02	1.7E-02	2.8E-02	4.6E-02
M 70 yr	3.2E-02	4.5E-02	6.4E-02	2.2E-02	3.5E-02	5.4E-02	1.6E-02	2.9E-02	4.8E-02

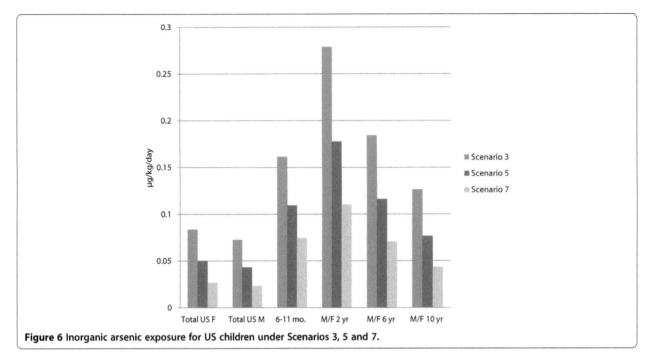

Figure 6 Inorganic arsenic exposure for US children under Scenarios 3, 5 and 7.

7 (50 percent terrestrial, 0 percent marine), inorganic arsenic represents 24 percent of the total arsenic exposure.

Food group contributions to inorganic arsenic exposure under Scenario 5 are presented in Figures 7, 8, and 9 for the general US population, 2 year-olds, and 6–11 month infants, respectively. For the general US population (Figure 7), 74 percent of the inorganic arsenic exposure came from just two food groups: grains, legumes and seeds (50 percent) and marine sources (24 percent). Grains, legumes and seeds also contributed 50 percent of the inorganic arsenic exposure for the 2 year-old population subgroup (Figure 8), but significant contributions were also seen from beverages (14 percent), marine sources (13 percent), and snacks and breakfast cereals

(12 percent). Food group contributions for the 6–11 month infant population subgroup (Figure 9) were dramatically different than for the 2 year-olds; the major contributor for 6–11 month infants was beverages (34 percent), followed by snacks and breakfast cereals (29 percent), and grains, legumes and seeds (25 percent).

Results obtained in this study can be compared to those found by researchers using different food consumption and food contamination databases. In terms of total arsenic exposure, our findings of a general US population average exposure of 0.16 µg/kg/day was considerably lower than the EPA estimate of 0.36 µg/kg/day (Xue et al. 2010) although our estimate for total arsenic exposure for the most exposed population subgroup (2 year-olds, 0.45 µg/kg/day) was fairly close to the EPA estimate of 0.54 µg/kg/day for ages 1 to 2. Differences in the findings are likely due to differences in the food consumption databases; our study used FDA TDS estimates for consumption of the various food items analyzed in the market basket survey using data derived from the 1994–1996 CSFII while the EPA used results from the 2003–2004 National Health Assessment and Nutritional Evaluation Survey. Both studies relied on analytical results obtained from the TDS although our study considered only data obtained between 2006 and 2008 while the EPA study used data collected from 1991 to 2004.

With respect to estimated exposures to inorganic arsenic, our findings are quite consistent with other studies conducted in the US. We estimated that inorganic arsenic levels ranged from 0.023 to 0.084 µg/kg/day for the general population, depending upon which scenario was used, and from 0.11 to 0.28 µg/kg/day for two year-

Table 4 Percentage of total arsenic represented as inorganic arsenic for selected population subgroups under all scenarios

Age/Sex subgroups	Inorganic arsenic contribution percentage (%)		
	6-11 mo.	M/F 2 yr.	M/F 6 yr.
Scenario 1 *100% T + 0% MS*	74	48	45
Scenario 2 *100% T + 10% MS*	77	54	50
Scenario 3 *100% T + 25% MS*	81	61	59
Scenario 4 *70% T + 0% MS*	52	34	31
Scenario 5 *70% T +10% MS*	55	39	37
Scenario 6 *70% T + 25% MS*	58	47	45
Scenario 7 *50% T + 0% MS*	37	24	22
Scenario 8 *50% T + 10% MS*	40	29	28
Scenario 9 *50% T + 25% MS*	44	37	36

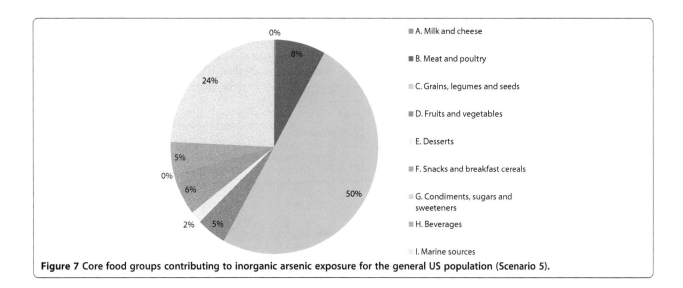

Figure 7 Core food groups contributing to inorganic arsenic exposure for the general US population (Scenario 5).

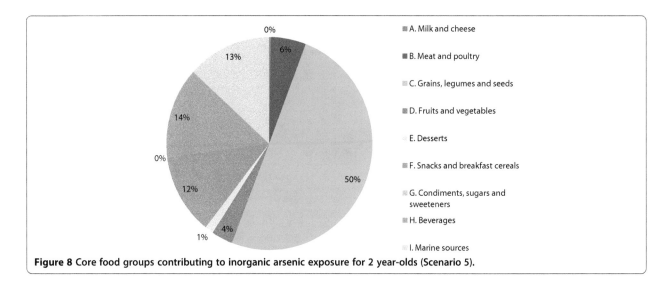

Figure 8 Core food groups contributing to inorganic arsenic exposure for 2 year-olds (Scenario 5).

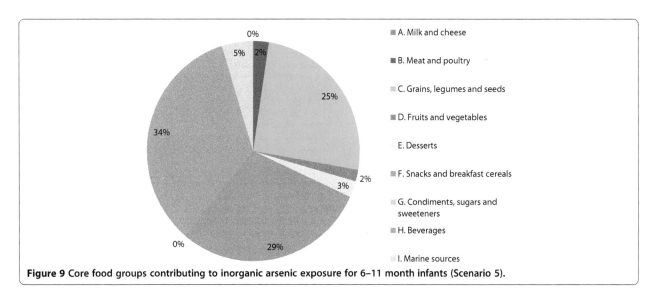

Figure 9 Core food groups contributing to inorganic arsenic exposure for 6–11 month infants (Scenario 5).

olds. The EPA's analysis, in which inorganic arsenic was considered to represent approximately 10 percent of the total arsenic, showed exposures ranging from 0.03 to 0.04 µg/kg/day for the general population and 0.08 to 0.23 µg/kg/day for children under five (Xue et al. 2010). The FDA's analysis, in which 10 percent of the total arsenic from marine sources and 100 percent of the total arsenic from terrestrial sources was considered to exist as inorganic arsenic, yielded inorganic arsenic exposures in the range of 0.026 to 0.25 µg/kg/day (Tao and Bolger 1999). All of the inorganic arsenic exposure estimates discussed above are significantly lower than those derived from EFSA, which range from 0.13 to 0.56 µg/kg/day for 2009 and 0.09 to 0.38 µg/kg/day for 2014 (EFSA 2009, 2014).

Conclusions

Using a variety of scenarios to estimate exposure to inorganic arsenic from foods for the general US population as well as for population subgroups, it was determined that inorganic arsenic exposure estimates for all population subgroups under all scenarios were at least slightly below the EPA's established RfD of 0.3 µg/kg/day and below the JEFCA health concern level of 2.0 to 7.0 µg/kg/day. Results compared closely with those from other US studies that used different food consumption databases, food contamination databases, and other methods to extrapolate inorganic arsenic levels from results of total arsenic analyses.

It is clear that accurate assessments of exposure to inorganic arsenic in the food supply require careful consideration of arsenic speciation among foods. Expressing arsenic exposure in terms of total arsenic and comparing such findings with the RfD for inorganic arsenic may lead to exaggerated exposures that imply a much greater risk than is actually present. In this paper, for example, the total arsenic exposure estimates for 2 year-olds (0.45 µg/kg/day) and for 6 year-olds (0.31 µg/kg/day) exceed the EPA's RfD for inorganic arsenic. Applying the speciation assumptions of Scenario 5 (10 percent of marine total arsenic is present as inorganic arsenic and 70 percent of terrestrial total arsenic is present as inorganic arsenic), for 2 year-olds and for 6 year-olds reduces inorganic arsenic exposure to 0.18 and 0.12 µg/kg/day, respectively.

While analysis of food samples for inorganic arsenic is generally much more expensive and difficult to perform than analysis for total arsenic, it may be possible to generate more accurate exposure assessments by focusing inorganic arsenic analyses on the primary food groups contributing to inorganic arsenic exposure. Grains, legumes and seeds, for example, contributed 50 percent of the estimated inorganic arsenic exposure under Scenario 5 for both the total US population and 2 year-olds, and 25 percent of the estimated inorganic exposure for 6–11

month infants. Focusing analytical approaches to detect inorganic arsenic from this food group could serve to reduce uncertainty in the exposure assessment by precluding the need to make assumptions concerning how much of the total arsenic from these foods is present in the inorganic form.

Abbreviations
EPA: U.S. Environment Protection Agency; IARC: International Agency for Research on Cancer; FDA: U.S Food and Drug Administration; RfD: Reference Dose; EFSA: European Food Safety Authority; JECFA: Joint FAO/WHO Expert Committee on Food Additives; PTWI: Provisional Tolerable Weekly Intake; TDS: Total Diet Study; CSFII: Continuing Survey of Food Intakes for Individuals; ppbm: Parts per billion.

Competing interests
The authors declare that they have no competing interests.

Authors' contributions
EJ was responsible for data collection and exposure assessment and assisted with preparation of the manuscript. CW assisted with the design of the analysis and with preparation of the manuscript. Both authors read and approved the final manuscript.

Acknowledgements
The authors thank to the Chilean Government and the National Commission for Scientific and Technological Research (CONICYT) for supporting this work through the Becas Chile Scholarships.

References
ATSDR Agency for Toxic Substances and Disease Registry (2005) Toxicological profile for arsenic. Available http://www.atsdr.cdc.gov/toxprofiles/tp2.pdf Accessed 27 Mar 2014

Carbonell-Barrachina A, Wu X, Ramírez-Gandolfo A, Norton G, Burló F, Deacon C, Meharg A (2012) Inorganic arsenic contents in rice-based infant foods from Spain, UK, China and USA. Environ Pollut 163:77–83

De Gieter M, Leermakers M, Van Ryssen R, Noyen J, Goeyens L, Baeyens W (2002) Total and toxic arsenic levels in North sea fish. Arch Environ Contam Toxicol 43(4):406–417

EFSA European Food Safety Authority (2014) Scientific Report of EFSA: Dietary exposure to inorganic arsenic in the European population. Available http://www.efsa.europa.eu/en/efsajournal/doc/3597.pdf Accessed 19 May 2014

EFSA European Food Safety Authority (2009) Scientific Opinion on Arsenic in Food. Available http://www.efsa.europa.eu/en/efsajournal/doc/1351.pdf Accessed 27 Mar 2014

Egan S, Bolger P, Carrington C (2007) Update of US FDA's total diet study food list and diets. J Expo Sci Environ Epidemiol 17:573–582

EPA Environmental Protection Agency (2011) Exposure Factors Handbook. Available http://www.epa.gov/ncea/efh/pdfs/efh-complete.pdf Accessed 27 Mar 2014

EPA Environmental Protection Agency (2012a) Integrated Risk Information System: Arsenic, inorganic. http://www.epa.gov/iris/subst/0278.htm Accessed 27 Mar 2014

EPA Environmental Protection Agency (2012b) Arsenic Rule. http://water.epa.gov/lawsregs/rulesregs/sdwa/arsenic/regulations.cfm Accessed 27 Mar 2014

EPA Environmental Protection Agency (2013) Arsenic in drinking water. http://water.epa.gov/lawsregs/rulesregs/sdwa/arsenic/index.cfm Accessed 27 Mar 2014

FAO/WHO Expert Committee on Food Additive (JECFA) (2011) Safety evaluation of certain contaminants in food. http://whqlibdoc.who.int/publications/2011/9789241660631_eng.pdf?ua=1 Accessed 19 May 2014

FDA Food and Drug Administration (2008) Total Diet Study. http://www.fda.gov/Food/FoodScienceResearch/TotalDietStudy/default.htm Accessed 27 Mar 14 2014

FDA Food and Drug Administration (2010) Total Diet Study Statistics on Element Results. http://www.fda.gov/downloads/food...totaldietstudy/ucm184301.pdf Accessed 21 May 2014

FDA Food and Drug Administration (2013a) Arsenic. http://www.fda.gov/Food/FoodbornelllnessContaminants/Metals/ucm280202.htm Accessed 27 Mar 2014

FDA Food and Drug Administration (2013b) FDA proposes action level for arsenic in apple juice. http://www.fda.gov/NewsEvents/Newsroom/PressAnnouncements/ucm360466.htm Accessed 27 Mar 2014

FDA Food and Drug Administration (2013c) Arsenic in Rice and Rice Products. http://www.fda.gov/Food/FoodborneIllnessContaminants/Metals/ucm319870.htm Accessed 27 Mar 2014

Georgopoulos P, Wang SW, Yang YC, Xue J, Zartarian V, McCurdy T, Ozcaynak H (2008) Biologically based modeling of multimedia, multipathway, multiroute population exposures to arsenic. J Expo Sci Environ Epidemiol 18(5):462–476

IARC International Agency for Research on Cancer (2012) Agents Classified by the IARC Monographs, Volumes 1–109. Available http://monographs.iarc.fr/ENG/Classification/ClassificationsAlphaOrder.pdf Accessed 27 Mar 2014

Lorenzana R, Yeow A, Colman J, Chappell L, Choudhury H (2009) Arsenic in seafood: speciation issues for human health risk assessment. Hum Ecol Risk Assess 15:185–200

Muñoz O, Devesa V, Suñer M, Velez D, Montoro R, Urieta I, Macho M, Jalon M (2000) Total and inorganic arsenic in fresh and processed fish products. J Agric Food Chem 48:4369–4376

Mohri T, Hisanaga A, Ishinishi N (1990) Arsenic intake and excretion by Japanese adults: a 7-day duplicate diet study. Food Chem Toxicol 28(7):521–529

Pennington J (1992) The 1990 revision of the FDA total diet study. J Nutr Educ 24(4):173–178

Sommella A, Deacon C, Norton G, Pigna M, Violante A, Meharg A (2013) Total arsenic, inorganic arsenic, and other elements concentrations in Italian rice grain varies with origin and type. Environ Pollut 181:38–43

Suñer M, Devesa V, Muñoz O, López F, Montoro R, Arias A, Blasco J (1999) Total and inorganic arsenic in the fauna of the Guadalquivir estuary: environmental and human health implications. Sci Total Environ 242:261–270

Suñer M, Devesa V, Clemente M, Vélez D, Montoro R, Urieta I, Jalón M, Macho M (2002) Organoarsenical species contents in fresh and processed seafood products. J Agric Food Chem 50:924–932

Tao S, Bolger M (1999) Dietary arsenic intakes in the United States: FDA total diet study, September 1991-December 1996. Food Addit Contam 16(11):465–472

Tavernise S (2013a) New Limits for Arsenic Proposed by FDA. http://www.nytimes.com/2013/07/12/health/new-limits-for-arsenic-proposed-by-fda.html Accessed 27 Mar 2014

Tavernise S (2013b) No Immediate Risk Is Found From Arsenic Levels in Rice. http://www.nytimes.com/2013/09/07/health/arsenic-levels-in-rice-products-not-a-health-risk-fda-says.html Accessed 27 Mar 2014

Valette-Silver N, Riedel G, Crecelius E, Windom H, Smith R, Dolvin S (1999) Elevated arsenic concentrations in bivalves from the southeast coasts of the USA. Mar Environ Res 48:311–333

WHO World Health Organization (2001) Arsenic and arsenic compounds. Environ Health Criteria 244: Available http://whqlibdoc.who.int/ehc/WHO_EHC_224.pdf Accessed 2 July 2014

Xue J, Zartarian V, Wang SW, Liu SV, Georgopoulos P (2010) Probabilistic modeling of dietary arsenic exposure and dose and evaluation with 2003–2004 NHANES data. Environ Health Perspect 118:345–350

Yost L, Schoof R, Aucoin R (1998) Intake of inorganic arsenic in the North American diet. Hum Ecol Risk Assess 4(1):137–152

5

Insights in agricultural practices and management systems linked to microbiological contamination of lettuce in conventional production systems in Southern Brazil

Sabrina Bartz[1], Claudia Titze Hessel[1], Rochele de Quadros Rodrigues[1], Anelise Possamai[1], Fabiana Oliveira Perini[1], Liesbeth Jacxsens[3], Mieke Uyttendaele[3], Renar João Bender[2] and Eduardo César Tondo[1*]

Abstract

Background: Three conventional lettuce farms were evaluated in Southern Brazil using a standardized self-assessment questionnaire with 69 indicators and a microbiological sampling plan in order to assess the status of current agricultural practices and management systems. The use of both tools aimed to identify the foremost contamination sources and control measures during the crop production. A total of 128 samples were taken (manure, soil, water, workers' hands and equipment, lettuce seedlings and lettuce heads) in four visits during the growth cycle of lettuces. Samples were analysed for hygiene indicators (*E. coli*) and presence of pathogens (*Salmonella* spp. and *E. coli* O157).

Results: Microbiological results indicated that *E. coli* counts were very low in all analysed samples and no pathogens were detected. These results could be explained partially because all farms had toilets near to the fields, they did not raise animals near the crops, fields were located in areas where flooding was not possible, they used organic fertilizers adequately composted, and irrigation water demonstrated good microbiological quality. The microbial results for manure and soil indicated that the composting time was of utmost importance to maintain minimal contamination levels for the duration of the cultivation period, as long as the quality of irrigation water was very important to prevent further contamination of the crop. On the other hand, the self-assessment questionnaire identified a moderate to high risk level concerning microbiological contamination in all evaluated farms, because they had no formal good agricultural practices implemented, technical support, water control, inspections, food safety registers or sampling plan for microbiological or chemical analyses.

Conclusion: These different results are important in order to provide information about the actual status of contamination (microbial sampling plan) and possible food safety problems in the future based on the results given by the questionnaire. Furthermore, the results of this study also highlighted the necessity to provide more safety during the fresh produce cultivation, being formal good agricultural practices implementation an important start to the fresh produce farms in Brazil, as well as to adopt a higher level of control activities in order to achieve lower risk levels.

Keywords: Conventional lettuce; Good agricultural practices; Microbiological contamination

* Correspondence: tondo@ufrgs.br
[1]Laboratório de Microbiologia e Controle de Alimentos, Instituto de Ciência e Tecnologia de Alimentos, Universidade Federal do Rio Grande do Sul (ICTA/UFRGS), Av. Bento Gonçalves, 9500, prédio 43212, Campus do Vale, Agronomia, Cep. 91501-970 Porto Alegre/RS, Brazil
Full list of author information is available at the end of the article

Background

Fresh produce is frequently associated with healthy diets because their nutritional properties and global production and consumption has increased significantly in the last years around the world (FAOSTAT, 2013; Warriner et al., 2009; Aruscavage et al., 2006). Intensive production systems and the lack of reliable good agricultural practices in the field are some of the reasons for the worldwide increasing numbers of foodborne illnesses associated to fresh produce (EFSA, 2014; Oilamat and Holley, 2012; Warriner et al., 2009; Beuchat, 2006; Sivapalasingam et al. 2004; Beuchat, 1996). Fresh produce can become contaminated with pathogens at any step of the supply chain, mostly due to natural, human or environmental factors (Olaimat and Holley, 2012; Oliveira et al., 2012; Itohan et al., 2011; Taban and Halkman, 2011). As a consequence, several foodborne outbreaks associated with leafy greens have been reported as primarily caused by *Salmonella* spp. and pathogenic *Escherichia coli* (Callejón et al., 2015; Buchholz et al., 2011; Warriner et al., 2009; Delaquis et al., 2007; Stine et al., 2005; Buck et al., 2003).

In Brazil, as in many other countries, lettuce (*Lactuca sativa* L.) is one of the most consumed leafy vegetables, attributable to year round availability, low cost and nutritional factors (Abreu et al., 2010; Mocelin and Figueiredo, 2009; WHO et al. 2008; Mattos et al., 2007). The Brazilian lettuce cultivation system is predominantly done in open fields, which are located for the most part at urban surroundings. Generally the distribution system occurs without refrigeration at any step of the postharvest chain, in contrast to practices in the European Union and United States, where cold chain and advanced logistics systems are applied (Brasil, 2013; Salla and Costa, 2012).

Food Safety Management Systems, for example, Good Agricultural Practices (GAP), at farm level are able to prevent and reduce bacterial contamination of fresh produce (Morgharbel and Masson, 2005; CDC, 2003; FDA, 1998). A number of factors has been identified as sources of microbial contamination, for example: organic fertilizers, soil, workers and equipment and, most noteworthy, water. Water has been identified as one of the most important sources of contamination of fresh produce. Irrigation waters and the fresh produce rinsing waters are recurrently used devoid of any disinfecting treatment (Rodrigues et al., 2014; Olaimat and Holley, 2012; Salem et al., 2011; Allende et al., 2008; Beuchat, 2006; Anderson et al., 1997).

Based on these evidences, the objective of the present study was to evaluate the status of current agricultural practices and management systems of conventional lettuce farms in the State of Rio Grande do Sul (RS), Southern Brazil, in order to identify major bottlenecks during the crop growing time related to conceivable microbiological contaminations. Insights were disclosed by combining microbiological analyses with the diagnosis of the risk level at farm circumstances, the status of implemented control measures and assurance activities and the system outputs at three typical Brazilian farms.

Methods
Characterization of the farms

Three family managed, smallholdings (approximately 2 to 3 hectares of land) in which lettuce was grown in a conventional production system were involved in the present study. Further on these production units were denominated farm 1, 2 and 3. These farms were chosen because they had typical characteristics of small farms were conventional lettuces and other leafy greens are cultivated in Brazil and also due to their similar conditions in terms of lettuce production. Before sampling collection, the owners were contacted and agreed to cooperate in the research. One of the farms was located in the rural area of Porto Alegre, the capital city of Rio Grande do Sul, the southernmost State of Brazil. The other two farms were located in the rural area of Viamão, a city neighboring Porto Alegre. Their cultivation system was in a open field.

The lettuce seedlings used to start off the plantations were delivered to the farms by different commercial suppliers. There were no formal good agricultural practices implemented or any other voluntary standard certified at the farms in the course of the sampling period. The fertilization procedures of the production fields were similar in all three farms. Organic fertilizers, over 90 days composted chicken manure, were purchased from local suppliers. None of the farms produced any kind of organic fertilizer.

The lettuce fields were irrigated by overhead sprinkler systems and the water was pumped from ponds located adjacent and at a lower level of the cultivation areas.

In all three farms the workers' households were located near the fields (less than 100 meters apart) and were equipped with toilets. Besides the intensive rainfall during the sampling period, flooding did not occur or affect the production fields. The farmers, during the sampling period, did not have cattle, poultry or other livestock animals in breeding process at their premises.

Microbiological sampling plan
Sampling locations and collection

A microbiological sampling plan was used with the intent of identifying contamination sources in the current agricultural practices. The sampling locations were selected based on literature review related to potential risk factors which may contribute to the microbiological contamination of lettuce. These locations were identified as critical sampling locations (CSL's), *i.e.*, sites in the production processes at which contamination, growth and/or survival of microorganisms may take place. In the present paper 12 CSL's were selected based on sources and potential risk factors of microbial contamination, starting from lettuce seedlings, soil and manure, irrigation and rinse waters, handlers, food contact equipment up to the final products (Rodrigues et al., 2014; Oilamat and Holley, 2012; Ilic et al., 2012).

The sampling period ranged from August to October 2012 and the microbial sampling plan was set up to obtain information about hygiene (*E. coli*) and safety levels (*Salmonella* spp., *E. coli* O157:H7). Samples of water, soil, manure, lettuce seedlings, lettuce heads, workers' hands and transport boxes were collected as previously described by Rodrigues et al. (2014).

All the samples were transported by car to the Laboratory of Microbiology and Food Control of the Institute of Food Science and Technology – ICTA/UFRGS inside thermal boxes. Analyses started in less than one hour after sampling.

Microbiological analyses

The analyses of microbiological parameters of each CSL are presented in Table 1. All the microbiological analyses were carried out according to Rodrigues et al. (2014).

Diagnostic instrument used to measure the food safety management systems

A questionnaire with 69 indicators was applied to gain insights into the level of the good agricultural practices and management system currently implemented on the farms, as previously described by Rodrigues et al. (2014). The questionnaires were answered by the farms' owners.

Weather conditions

Temperature and cumulative precipitation of the week prior to sampling and including the sampling day (8 days) were obtained from the National Institute for Meteorology of Brazil (Instituto Nacional de Meteorologia (INMET), http://www.inmet.gov.br/portal/). Table 2 shows the averages of temperature and precipitation during the sampling period.

Statistical analyses

Statistical analyses were performed with SPSS Statistics version 21 at $p < 0.050$. Bivariate correlations between the indicators were determined by calculating the Spearman's Rho coefficient using the raw enumeration data. Kruskal-Wallis or Mann-Whitney U tests were used to evaluate the influence of different factors. Pair wise tests were performed to identify the significant differences between individual categories when significant differences were found. In case of 'n' pair wise comparisons,

Dunn-Sidak correction was applied, resulting in adjusted individual p' values: $p' = 1-(1-p)^{1/n}$, in which $p = 0.050$ to obtain a family-wise error rate of 5%.

Results

Microbiological contamination

The presence of *E. coli* in the collected samples from manure, manured soil before setting the lettuce plantlets into the field and soil along the growth cycle of the lettuce crops presented mostly counts below the detection limits (Table 3). The highest count of *E. coli* (2.00 $\log_{10}$ CFU/g) was observed in two samples: one sample of manure and another of soil (Table 3). There was no significant difference in *E. coli* counts between manured soil and soil samples for the duration of the sampling period (Kendall's tau-c, p = 0.803). There were no significantly differences in *E. coli* counts in manure among farms (Kruskal-Wallis Test, p = 0.368). *Salmonella* spp. and *E. coli* O157:H7 were not found in any sample. The *E. coli* concentration along the growth cycle in manure, manured soil and soil in the three farms is demonstrated in Figure 1.

Lettuce seedlings were collected only at the time of planting the seedlings in the field. *E. coli* counts ranged from <1.00 $\log_{10}$ CFU/g to 2.30 $\log_{10}$ CFU/g (average of 1.43 ± 0.75 $\log_{10}$ CFU/g). The highest count was observed on seedlings at farm 1. During the growth cycle of the lettuces, the *E. coli* distribution was similar (Kruskal-Wallis Test, p = 0.560) (Figure 2). However, the highest *E. coli* counts were observed two and one week before harvest. At harvest, all *E. coli* counts were below the detection limit (Figure 2). *E. coli* counts were similar on the lettuce head samples collected at all farms (Kruskal- Wallis Test, p = 0.162), ranging from <1.00 ± 0.00 $\log_{10}$ CFU/g to 1.12 ± 0.14 $\log_{10}$ CFU/g. The rinsed lettuce heads presented *E. coli* counts below the detection limits and no pathogens were found on any sample of seedlings and lettuces.

Water samples collected from ponds, sprinklers and rinsing tanks presented low counts of *E. coli* and 88.5% of the samples counts were below the detection limit (Table 3). Counts of positive samples ranged from 1 to 1.4 $\log_{10}$ MPN/100 ml. No statistical differences were determined for *E. coli* among the three water sources during the growth cycle of the crop (Kruskal- Wallis Test, p = 0.739). No pathogens were detected in any analyzed sample. During the lettuce growth cycle, the distribution of *E. coli* showed no significant differences among farms and time of sampling (Kruskal- Wallis Test, p = 0.212). No pathogens were found in any water sample.

The samples of the transport boxes and workers' hands of the three farms were collected only at harvest. All samples showed *E. coli* counts below the detection limit (Table 3).

Table 1 Description of Critical Sampling Location (CSLs), samples, periodicity, microbiological parameters, methodologies, results interpretation and references

CSL	Description	Samples	Time	Microbiological parameters	Methodology	Interpretation of the results*	References
1	Manure	3 samples	T0	*E. coli*	ISO 21528-2:2004	10^3 cfu/g	MAPA/ IN n°46. (2011)
				E. coli O157:H7	ISO 16654:2001	A/25g	ND
				Salmonella spp.	ISO 6579:2002	A/25g	MAPA/ IN n°46. (2011)
2	Manured soil	3 samples → 3 x 3 pooled	T0	*E. coli*	ISO 21528-2:2004	10^3 cfu/g	MAPA/ IN n°46. (2011)
				E. coli O157:H7	ISO 16654:2001	A/25g	ND
				Salmonella spp.	ISO 6579:2002	A/25g	MAPA/ IN n°46. (2011)
3	Soil	3 samples → 3 x 3 pooled	T1	*E. coli*	ISO 21528-2:2004	10^3 cfu/g	MAPA/ IN n°46. (2011)
			T2	*E. coli* O157:H7	ISO 16654:2001	A/25g	ND
			T3	*Salmonella* spp.	ISO 6579:2002	A/25g	MAPA/ IN n°46. (2011)
4	Seedlings in soil	1 sample → 1 x 3 pooled	T0	*E. coli*	ISO 21528-2:2004	10^2 cfu/g	RDC n°12 (2001)
				E. coli O157:H7	ISO 16654:2001	A/25g	ND
				Salmonella spp.	ISO 6579:2002	A/25g	RDC n°12 (2001)
5	Seedling	1 sample	T0	*E. coli*	ISO 21528-2:2004	10^2 cfu/g	RDC n°12 (2001)
6	Lettuce	3 samples → 3 x 3 pooled	T1	*E. coli*	ISO 21528-2:2004	10^2 cfu/g	RDC n°12 (2001)
			T2	*E. coli* O157:H7	ISO 16654:2001	A/25g	ND
			T3	*Salmonella* spp.	ISO 6579:2002	A/25g	RDC n°12 (2001)
7	Lettuce after washing	3 samples → 3 x 3 pooled	T3	*E. coli*	ISO 21528-2:2004	10^2 cfu/g	RDC n°12 (2001)
				E. coli O157:H7	ISO 16654:2001	A/25g	ND
				Salmonella spp.	ISO 6579:2002	A/25g	RDC n°12 (2001)
8	Rinse water	100 ml	T3	*E. coli*	20 TH APHA (1998)	2×10^2 MPN/100ml	CONAMA. n°357 de 2005
				E. coli O157:H7	ISO 16654:2001	A/25ml	ND
				Salmonella spp.	ISO 6579:2002	A/25ml	ND
9	Irrigation water source	100 ml	T0 T1	*E. coli*	20 TH APHA (1998)	2×10^2 MPN/100ml	CONAMA. n°357 de 2005
			T2	*E. coli* O157:H7	ISO 16654:2001	A/25ml	ND
			T3	*Salmonella* spp.	ISO 6579:2002	A/25ml	ND
10	Irrigation water from tap	100 ml	T0 T1	*E. coli*	20 TH APHA (1998)	2×10^2 MPN/100ml	CONAMA. n°357 de 2005
			T2	*E. coli* O157:H7	ISO 16654:2001	A/25ml	ND
			T3	*Salmonella* spp.	ISO 6579:2002	A/25ml	ND
11	Swab of farmers' hands	3 x 25 cm²	T3	*E. coli*	ISO 21528-2:2004 and AOAC (1998)	≤ 0.7 log cfu/25 cm² (below detection)	Jacxsens. et al. (2010)
12	Swab of transport boxes of lettuce	3 x 50 cm²	T3	*E. coli*	ISO 21528-2:2004	≤ 0.7 log cfu/25 cm² (below detection)	Jacxsens. et al. (2010)

A: absent; ND: not defined by official regulation.
T0: At planting. T1: Two weeks before harvest. T2: One week before harvest. T3: At harvest.

Weather parameters

Regarding weather parameters (temperature and precipitation), results were significantly different (Kruskal-Wallis Test, p < 0.001) among the farms and the sampling days throughout the sampling period (Table 2). At farm 1, on the first day of sampling (T0), the highest count of *E. coli* found on soil seedling samples was 2.30 $\log_{10}$ CFU/g. On that day the amount of rain fall was, statistically, the lowest in comparison to the other sampling days (Mann-Whitney *U* Test, p < 0.001) (Table 2; Figure 1). On

the other farms, no statistical differences were observed both for *E. coli* counts and rain fall volumes during the sampling period.

Temperature at transplanting day was similar to temperatures observed at two and one week before harvest (Mann-Whitney *U* Test, p = 0.446, p = 0.64, respectively) and significantly different from the harvest day (Mann-Whitney *U* Test, p = 0.002). Between the sampling periods of one and two weeks before harvest, temperatures were as well significantly different (Mann-Whitney *U* Test,

Table 2 Mean and standard deviation of temperature and precipitation during sampling period in three farms producing conventional lettuces in Southern Brazil

Farm	Visit	Temperature* (°C)	Precipitation* (mm)
1	T0	18.01 ± 2.58^a	0.58 ± 1.31^a
	T1	19.02 ± 2.46^b	4.84 ± 12.80^b
	T2	19.24 ± 1.96^c	9.65 ± 13.97^c
	T3	17.71 ± 2.33^d	23.38 ± 35.06^d
2	T0	19.02 ± 2.46^a	4.84 ± 12.80^a
	T1	16.92 ± 3.15^b	1.49 ± 3.71^b
	T2	21.70 ± 1.99^c	5.66 ± 6.93^c
	T3	19.40 ± 1.90^d	3.70 ± 5.24^d
3	T0	19.02 ± 2.46^a	4.84 ± 12.80^a
	T1	16.92 ± 3.15^b	1.49 ± 3.71^b
	T2	21.70 ± 1.99^c	5.66 ± 6.93^c
	T3	19.40 ± 1.90^d	3.70 ± 5.24^d

a,b,c,d : Different letters indicate statistically significant differences between the different sampling period.

p = 0.004), however no significant difference was observed in *E. coli* counts on samples.

The rain fall amounts were similar between the transplanting day, one week before harvest and at harvest (Mann-Whitney *U* Test, p = 0.064 and p = 0.426, respectively). However, two weeks before harvest the amount of rain fall was statistically higher when compared to one week before and at harvest (Mann-Whitney *U* Test, p < 0.01 for both), but the *E. coli* counts remained similar.

Diagnosis of the current good agricultural practices and management system

The context of the farmers appraised revealed that the conventional lettuce farms had a high risk context towards microbiological safety and crop hygiene. The calculated averages for product and process characteristics reached an index of 3.0 for all the three farms, because they have similar products and production practices (Table 4).

Indicators of organization & chain processing scored 2.46 (farm 1), 2.69 (farm 2) and 2.54 (farm 3), indicating moderate to high level of risk (Table 4). The riskiness of the organization of the farms was very similar, except for the indicators 'technical staff of the farm' and 'variability in workforce'. Farm 1 had a stable workforce and additionally technological insights were as well present. At farm 2 also a good technological staff was present, but the activities had to rely on part time working personnel. For farm 3 the situation was rather the opposite. Working personnel at the premises was already an effective and a stable workforce for a long period of time. Nonetheless, the technological knowledge was not present. The indicators at level 2 (moderate risk) were 'extent of

power in supplier relationships' and 'logistic facilities' for all three farms. However, all the other indicators were classified as at high risk level (level 3) for the three farms (sufficiency of operator competences, extent of management commitment, degree of employee involvement, level of formalization, sufficiency supporting information systems, food safety information exchange, and inspections of food safety authorities).

The indicated levels of the control activities in the good agricultural practices of the farms are specified in Table 4. The mean score of the design or set-up of control activities was 1.53. An indication that these activities were absent (level 1) or conducted on a basic level, using historical and common knowledge (level 2), and no sector information or information from suppliers was applied (level 3), nor tailored to the farms own situation (level 4).

The profiles were very similar for all the three farms, though farm 3 differs from farms 1 and 2 on 'partial physical intervention' (rinsing step), because rinsing of the lettuce crops was not conducted at farm 3. Farms were operating mainly at basic level (level 1) with regards to items related to 'equipment hygienic design maintenance program', 'sanitation program', 'packaging equipment', 'water control', 'sampling for microorganisms', 'analyzing methods for pathogens' and 'corrective actions'. An indication that all these control activities were not in place on the three farms (Table 4).

The indicators 'storage facilities', 'personal hygiene', 'raw materials control', 'fertilizer program', 'irrigation method' were classified at level 2. That level suggests that these activities were performed based on the knowledge of the farmers and not based on inputs from guidelines, sector organizations or government (Table 4).

For the farms at which rinsing of the lettuce heads was implemented after harvest (farms 1 and 2), the rinsing was also done based on their individual knowledge. Supplier control of the seedlings and manure composting were well achieved (level 3, best situation) because all farms bought seedlings from the same supplier and fertilizers had been already composted over 90 days before arrival to the farms.

Moreover, the actual operation of control activities was lower (averages of 1.43 for all three farms – Table 4) compared to the design or set-up of the control measures. This situation is indicating that the control measures were not implemented and applied in practice. Only the indicator 'compliance to producers' received a level 2, because the growers comply to their own working method.

Also assurance activities such as 'translation of stakeholder requirements', 'use of feedback information', 'validation activities' and 'verification activities', 'documentation system' and 'record keeping' were not present or had

Table 3 Sampling location, sample type, number of samples and results for microbiological analysis

| | | | Hygiene indicators | | | | Pathogen indicator | |
| | | | E. coli (mean and stdv) | Number of samples per E. coli counts | | | Salmonella | E. coli O157:H7 |
CSL	Sample	n		<1.0 log	≥1.0 and < 2.0 log	≥2.0 and < 3.0 log	A/P*	A/P*
1	Manure	9	1.11 ± 0.33 cfu/g	8	0	1	A	A
2	Manured soil	9	<1.00 ± 0.00 cfu/g	9	0	0	A	A
3	Soil	27	1.05 ± 0.20 cfu/g	23	3	1	A	A
4	Seedlings in soil	3	1.43 ± 0.75 cfu/g	2	0	1	A	A
5	Seedlings	3	1.00 ± 0.00 cfu/g	2	1	0	A	A
6	Lettuce	27	1.06 ± 0.22 cfu/g	23	3	1	A	A
7	Lettuce after washing	6	1.00 ± 0.00 cfu/g	5	1	0	A	A
8	Rinse water	2	1.00 ± 0.00 MPN/100 ml	2	0	0	A	A
9	Irrigation water source	12	1.03 ± 0.012 MPN/100 ml	10	2	0	A	A
10	Irrigation water from tap	12	1.04 ± 0.12 MPN/100 ml	10	1	1	A	A
11	Swab of farmers' hands	9	1.00 ± 0.00 cfu/25 cm^2	9	0	0	-	-
12	Swab of transport boxes of lettuce	9	1.00 ± 0.01 cfu/25 cm^2	9	0	0	-	-
Total		128		112	11	5	-	-

* A: absent in 25 g or 25 ml; P: presence in 25 g or 25 ml; stdv: standard deviation.

not been yet developed. An indication that the farms could not demonstrate that they were working correctly (mean level of 1 for all).

The system output of the current good practices for the conventional lettuce farms was also low (mean 1 for all the three farms). The reason for this was that no information was available about the system output: no inspection or audit was performed, no samples (for microbiological or chemical analyses) were taken, no visual quality was evaluated, and no non-conformities were recorded or evaluated. Consequently no actual evaluation of the system output could be completed (Table 4).

Discussion

In the present study low levels of microbiological contamination were found in samples collected from small farms producing conventional lettuces in Southern Brazil, even though a high risk context towards microbiological safety and crop hygiene was verified in all of them based on the self-assessment questionnaire. These different results may indicate that some good agricultural practices were in place, however no formal control was applied.

For example, low levels of contamination and the absence of *Salmonella* spp. and *E. coli* O157:H7 observed in manure were attributed to the fact that all farms purchased manure from commercial suppliers, which was already composted for over 90 days. Several authors described that adequate composting time will effectively reduce contamination (Oliveira et al., 2012; Fischer-Arndt et al., 2010; James, 2006; Millner, 2003; MAFF, 2000) and particular pathogens like *E. coli* and *Salmonella* spp. can

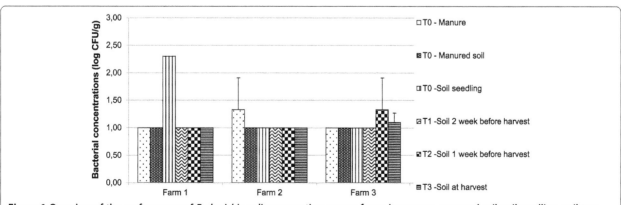

Figure 1 Overview of the performance of *Escherichia coli* enumeration among farms in manure, manured soil, soil seedling, soil two and one week before harvest and at harvest.

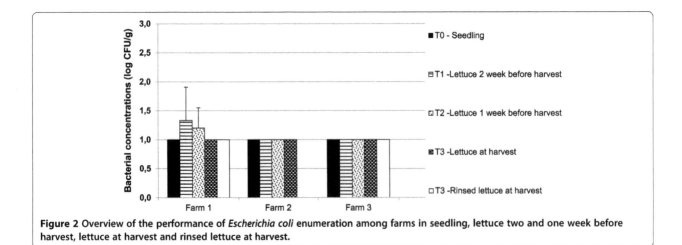

Figure 2 Overview of the performance of *Escherichia coli* enumeration among farms in seedling, lettuce two and one week before harvest, lettuce at harvest and rinsed lettuce at harvest.

survive at maximum to 90 days in soil as well as in manure (Heaton and Jones, 2007; Nicholson et al., 2005).

Moreover, the composted manures used at farms were added to the soil at least two weeks prior setting the seedlings into the field and after that no more manure was applied to supply nutrients to the lettuce plants. Also in the evaluation of the current good agricultural practices related to manure management, indicator 'organic fertilizer program', a moderate level 2 was given for the three farms, indicating that they used and manipulated manure based on generic knowledge from their suppliers (Table 4).

At planting and at harvest, all *E. coli* counts were below the detection limit (<1.00 $\log_{10}$ CFU/g), demonstrating good quality of lettuce seedlings and final product (lettuce) in attendance to the parameters of the Brazilian legislation (Brasil, 2001) that sets 10^2 CFU/g as the maximum acceptable limit for *E. coli* counts. The fact that no *E. coli* was detected on lettuces can be attributed to the low pressure of *E. coli* in the manure, manured soil around the crop and low contamination of the irrigation water. Corroborating these results, EFSA (2014) reported that several reasons can be attributed to the variation in *E. coli* numbers on leafy greens and the relationship between primary production practices and numbers of *E. coli* in final product is very variable. Even though, it is difficult to define which is the main cause of this variation, the microbial quality of manure and irrigation water are frequently cited (EFSA, 2014).

In the present study, the water supply was considered a high risk (Table 4), especially because the water came from ponds (Richardson et al., 2009) and there was no further treatment, however water sampled from ponds, sprinklers and rinsing tanks presented low levels of contamination by *E. coli*. All analyzed samples were in accordance with the Brazilian regulation for irrigation of vegetables (CONAMA, 2005), which establishes a limit of 2×10^2 CFU/100 ml for thermotolerant coliforms.

Similarly, no *Salmonella* spp. or *E. coli* O157:H7 were isolated from any of the analysed water samples. In a different study conducted in organic farms of the same region of Brazil (Rodrigues et al., 2014), the presence of *Salmonella* spp. and *E. coli* O157:H7 was detected in two samples (irrigation and rinsing tank water), after a flooding event. It is important to mention that in the farms investigated in the present study, the water supply (ponds) and the crop fields were located in elevated areas were flooding could not occur. Other authors observed the influence of flooding in the variation of pathogens levels (Liu et al., 2013, Castro-Ibañez et al., 2013; Cevallos-Cevallos et al., 2012; Tirado et al., 2010; Franz et al., 2005; Girardin et al., 2005; Rose et al., 2001).

Water has been identified as the source of microbial contamination of several foodborne outbreaks involving leafy vegetables around the world (Itohan et al., 2011; Delaquis et al., 2007; Beuchat, 1996). Pathogenic bacteria such as *E. coli* O157:H7 are often associated with outbreaks of waterborne diseases, resulting from inadequate treatments of the water used for irrigation and rinsing of fruits and vegetables (Levantesi et al., 2012; Moyne et al., 2011). Furthermore, in the present study, farms 1 and 2 used the same irrigation water source to rinse the lettuces after harvest. At farm 3 no rinsing of the lettuce heads did take place. The results indicated that no significant differences were observed for *E. coli* counts before, after or without the rinsing procedure, even though the water supply was considered a high risk of contamination (Table 4) and there was no water control.

No pathogens were identified in any crop sample and no increases in the microbial counts were as well observed after the rinsing process, demonstrating just the opposite in our study of what was ascertained in a study conducted by Antunes (2009).

Regarding organization and chain characteristics (Table 4), the technological staff present in farm 1 had received technical support provided by local government (city), while in

Table 4 Scores and calculated mean attributed to the indicators of food safety management system

Indicators	Farm1	Farm 2	Farm 3	Description of situation
I. Context factors (overall)[a]				
Product and process characteristics				
Risk of raw materials microbial	3	3	3	Seedlings and manure purchased from commercial suppliers without any Good Agricultural Practice implemented. Irrigation water without any treatment. Seedlings in direct contact with soil.
Risk of final product microbial	3	3	3	The lettuces crops growing in direct contact with soil and without covering.
Production system	3	3	3	Open cultivation field and contact with soil.
Climate conditions	3	3	3	The farms were located in subtropical areas, with uncontrolled climate conditions.
Water supply	3	3	3	All producers used water from ponds, without treatment.
Mean product and process	*3,00*	*3,00*	*3,00*	
Organization and chain				
Presence of technological staff	2	3	3	Farm 1 had technical support provided by government department (of the city). Farm 2 and 3 had no technical support.
Variability in workforce composition	1	3	1	Farm 2 had a high turnover of employees and temporary operators were commonly used. Farm 1 and 3 had low turnover, with occasonaly temporary operators.
Sufficiency of operator competences	3	3	3	Operators with no training in food safety control, only practice experience in the field.
Extent of management commitment	3	3	3	All three farms had no written food safety policy and no official quality team.
Degree of employee involvement	3	3	3	There was no safety control sistems implemented in the farms.
Level of formalization	3	3	3	No meetings sistem implemented for instructions communication exist in all producers.
Sufficiency supporting information systems	3	3	3	None of the producers had standard information system for food safety control decisions.
Severity of stakeholders Requirements of	3	3	3	Steakholders did not ask for any QA requirements.
Extent of power in supplier relationships	2	2	2	All farms required from their manure suppliers to compost the manure as a prerequisit for purchase.
Food safety information exchange	3	3	3	No sistematic exchange of information on food safety issues were done with the suppliers of the three producers.
Logistic facilities	2	2	2	Transport of the final products to the distributer done by trucks in protected conditions (covered) but room temperature.
Inspections of food safety authorities	3	3	3	Never a inspection were done in the three farms.
Supply source of initial materials	1	1	1	Only local suppliers of major initial materials
Mean organisation and chain	*2,46*	*2,69*	*2,54*	
II. Control activities design[b]				
Hygienic design of equipment and facilities	1	1	1	None specific hygienic design required for equipement and facilities among the producers.
Maintenance and calibration program	1	1	1	No manteinance and calibration program apllied in any of the producers.
Storage facilities	2	2	2	Storage was made in ambient conditions in all farms.
Sanitation program(s)	1	1	1	The producers had no specific sanitation program implemented.

Table 4 Scores and calculated mean attributed to the indicators of food safety management system (Continued)

Personal hygiene requirements	2	2	2	No specific hygiene instructions were followed by the operators but washing facilities and toillets were available next to the field in all farms.
Incoming material control	2	2	2	Incoming material control was done by visual inspections based on historical experience in all farms.
Packaging equipment	2	2	2	Use of non specific plastic boxes to pack the lettuce.
Supplier control	2	2	2	The farms had no specific pre requisites for supplier selection.
Organic fertilizer program	2	2	2	Pre composted manure purchased from local suppliers in all producers.
Water control	1	1	1	There was no water control in all farms.
Irrigation method	2	2	2	All producers used sprinkler as the irrigation method.
Partial physical intervention	2	2	1	General partial physical intervention applied by washing the lettuce and external leaves removed
Analytical methods to assess pathogens	1	1	1	The presence of pathogens were never analyzed by any of the producers.
Sampling plan for microbial assessment	1	1	1	The producers had no sampling plan implemented.
Corrective actions	1	1	1	The farms had no corrective actions described.
Mean control activities design	*1,53*	*1,53*	*1,53*	

III. Control activities operation[b]

Actual availability of procedures	1	1	1	The procedures were not documented in all the three farms.
The actual of compliance to procedures	2	2	2	The operators executed tasks according to their own experience and ad-hoc basis.
Actual hygienic performance of equipment and facilities	1	1	1	The hygienic design is not considered to be important for food safety.
Actual storage/cooling capacity	1	1	1	The farms had no cooling storage facility available.
Actual process capability of partial physical intervention	2	2	2	The partial physical intervention were done without standard parameters and no control charts.
Actual process capability of packaging	2	2	2	Packaging were done without regular parameters and based on the lettuce size.
Actual performance of analytical equipment	1	1	1	No analytical analyses were done in all farms.
Mean control Activities operation	*1,43*	*1,43*	*1,43*	

IV. Assurance activities[b]

Translation of stakeholder requirements into own HSMS requirements	1	1	1	Stakeholder requirements were not present in all three farms.
The systematic use of feedback information to modify HSMS	1	1	1	The farms had no HSMS implemented.
Validation of preventive measures	1	1	1	The producers had no preventive measures implemented and validated.
Validation of intervention processes	1	1	1	Intervention processes have never been validated and were done based on their own knowledge.
Verification of people related performance	1	1	1	The producers had no documented procedures described, so no verification was done.
Verification of equipment and methods related performance	1	1	1	No procedures of verification for equipment and methods were preformed in all producers.
Documentation system	1	1	1	Documentation were not available in all the farms.
Record keeping system	1	1	1	no record keeping system were present in all three farms.
Mean assurance activities	*1,00*	*1,00*	*1,00*	

Food safety management system Output[c]

Food safety Management System evaluation	1	1	1	No inspection or audit of the Food Safety Management System were done in all produceres.

Table 4 Scores and calculated mean attributed to the indicators of food safety management system (Continued)

Seriousness of remarks of remarks	1	1	1	Audits on HSMS were never performed.
Hygiene related and microbiological food safety	1	1	1	No records of hygine related and microbiological food safety complains were available in the farms.
Chemical safety complaints of customers	1	1	1	Chemical complains records were not avalilable in the producers.
Typify the visual quality complaints	1	1	1	No records about quality complaints were available in the farms.
Product sampling microbiological performance	1	1	1	The microbiological performance is not known once no microbiological analyses were done on regular basis.
Judgment criteria microbiological	1	1	1	Microbiological analyses were not performed in the farms.
Non conformities	1	1	1	The performance of the HSMS was not possible once no coformities registration were available
Mean food safety output	*1,00*	*1,00*	*1,00*	

[a]I Context factors: product and process characteristics and organization and chain characteristics were evaluated based on three risk levels: level 1 (low risk); level [2](medium risk); and level 3 (high risk).

[b]II Control activities design: evaluates the designs of control activities; III evaluates the actual operation or implementation of control activities; IV evaluates the assurance activities in good agricultural practices based on four levels: level 1 (non-existing or not implemented); level 2 (activities done at basic level based on own knowledges and historical information); level 3 (activities implemented based on sector information or guidelines); level 4 (activities adapted and tailored to the specific situation on the farm).

[c]IV Food safety system output indicators: evaluation based on external or governmental audits, records, microbial and chemical analysis: level 1 (not done or no information available); level 2 (limited information available); level 3 (more systematic information is available); level 4 (systematic informations available and good results are obtained).

farms 2 and 3 no technical support was given. At the same time, farm 1 and 3 had a stable workforce, while farm 2 demonstrated a high turnover. Some authors described that the stability of the workforce can help the companies to prevent food safety questions and problems (Kirezieva et al., 2013a; Luning et al., 2011). At the same time the other organization characteristics demonstrated that all farms were operating in a very low level of organization, what is common in family based companies (Lunning et al, 2011; Powell et al., 2011), with the operators without any king of food safety training, no safety control systems implemented or written, no standard information about safety control systems, stakeholders without any quality assurance required, transport of the final product without temperature control and no inspection done by official authorities. It is well known that a trained workforce can help the companies to implement the good agricultural practices, once the employees know their responsibilities with the food safety issues (Kirezieva et al., 2013b) and that governmental inspections are also important to assure the compliance of the companies with the good practices (Jafee and Masakure, 2005; Kierzieva et al., Kirezieva et al. 2013a). It has also been demonstrated that the practice of keeping registration and documents in the primary production level is not usual in other countries (Jevsnik et al., 2008; Nieto-Montenegro et al., 2008), however, this could be a good procedure to be implemented in Brazilian farms in order to reach higher food safety levels.

It might be assumed that the studied conventional lettuce farms were in a moderate to high level of risk in microbiological contamination due to product and process characteristics (Kirezieva et al., 2013b), once the seedling where purchased from commercial suppliers without formal good agricultural practices implemented, lettuce crops were in direct contact with soil, farmers located in subtropical areas without climate conditions control, there was no treatment of irrigation water, and the cultivation was in open fields (Table 4). That context level found in the three farms suggests that a medium to advanced level of good agricultural practices and management system should be present in order to have a good system output as described by authors such as Osés et al. (2012) and Kirezieva et al. (2013b). However, the good practices and management of all investigated farms were informal and very basic, which may implicate in a high risk of food safety problems (Uyttendaele et al., 2014). Moreover, in the conventional lettuce farms investigated, there was no system output because of the lack of registered information and controls. This results could be explained because in Brazil there is no governmental requirement for that and producers are not stimulated to make quality records. A similar situation was observed in organic farms in the same region of Brazil (Rodrigues et al., 2014). Different circumstances was reported by Kirezieva et al. (2015) for companies located in the European Union where lower to moderate risk of production and supply chain context was found because, among other factors, controlled water sources were used and the cultivation was done in a protected area. The microbial load and pressure in the conventional farms analysed in the present study were lower compared to the samples collected in organic farms studied by Rodrigues et al. (2014), who reported higher *E. coli* counts and also the presence of *Salmonella* and *E. coli* O157:H7.

The major differences between conventional and organic farms studied in Southern Brazil were the manure and composting of manure, which was conducted by the organic farms themselves with uncontrolled manner while a good manure management and control was evaluated for the conventional farms. Also, no animals were present on conventional farms what may contributed in the reduction of *E. coli* pressure on the water sources. Furthermore, a good water quality was verified in conventional farms, what was not the case in the organic ones (Rodrigues et al., 2014).

Conclusions

The use of the risk based sampling plan in combination with the diagnostic questionnaire allowed to analyse the microbiological aspects and the status of management systems of conventional lettuce farms in Southern Brazil.

Although all farms had no formal good agricultural practices implemented and there was no technical support in any of them, the microbial parameters showed very low levels of contamination, including the final products (lettuce heads). These results are plausible for the reason that Brazilian regulatory bodies do not enforce the implementation of good agricultural practices, nonetheless farmers are frequently aware that farm organization and hygienic procedures are necessary in order to maintain food safety and good productive levels. As an example, all analyzed farms had toilets near to the fields, providing adequate personnel hygienic practices. Further, the farms did not raise animals such as cows, pigs and hens, ultimate sources of cross contamination of the fields, remarkably, as a consequence of rain falls. In addition, the fields were located in areas where flooding was not possible. Another important aspect to take into account, concerning the organic fertilizer that was appropriately composted, not impacting on the contamination of the crops. Similarly, the good quality of the irrigation waters used, evidenced by the microbial analyses, did not influence the contamination of the final product.

Good practices should be applied during all food chain, farm to fork. It has been observed that in the last years, outbreaks caused by fresh produce are increasing around the world, suggesting that, in that particular step of the chain, primary production, more efforts are needed in order to get more safety.

Even though the fact that all the microbial results were very low and no pathogen was determined in any of the analysed samples, attention should be given to the results of the self-assessment questionnaire that indicated moderate to high risk levels at all farms. These different results are important in order to provide information about the actual status of contamination (microbial sampling plan) and possible food safety problems in the future based on the results given by the questionnaire. Furthermore, the results of this study also highlighted the necessity to provide more safety during the fresh produce cultivation, based on the bottlenecks identified by the self-assessment questionnaire, being formal good agricultural practices implementation an important start to the fresh produce farms in Brazil, as well as to adopt a higher level of control activities in order to achieve lower risk levels.

Competing interests

The authors declare that they have no competing interests.

Authors' contributions

SB carried out initial contact with producers, performed sampling collection, carried out the microbial analyses, interviewed the producers with the self-assessment questionnaire, elaborated critical analyses based on results, drafted the manuscript. Carried out revisions on manuscript. CTH - Helped with the microbial and statistical analyses, participated in scientific discussions. RQ - Helped with the sample collection. AP - Prepared sampling material and contributed with for microbial analyses, participated in scientific discussions. FP - Prepared sampling material and contributed with for microbial analyses, participated in scientific discussions. LJ - Planned the sampling collection and general organization of experiments. Helped with the interpretation of the self-assessment questionnaire results and discussion. Participated in scientific discussions. MU - Planned the sampling collection and general organization of experiments. participated in scientific discussions. RJB - Revised the manuscript and added inputs. ECT organised research team in all activities of the manuscript. Planned the sampling collection and general organization of experiments. participated in scientific discussions. Contributed with laboratory infra-estructure. Participated in elaboration of the manuscript. All authors read and approved the final manuscript.

Acknowledgements

This research has been supported by the European Community's Seventh Framework Program (FP7) under grant agreement no. 244994 (project VEG-i-TRADE).

Author details

[1]Laboratório de Microbiologia e Controle de Alimentos, Instituto de Ciência e Tecnologia de Alimentos, Universidade Federal do Rio Grande do Sul (ICTA/UFRGS), Av. Bento Gonçalves, 9500, prédio 43212, Campus do Vale, Agronomia, Cep. 91501-970 Porto Alegre/RS, Brazil. [2]Laboratório de Pós-Colheita, Faculdade de Agronomia, Universidade Federal do Rio Grande do Sul, Av Bento Gonçalves, 7712. 91540-000 Porto, Alegre/RS, Brazil. [3]Department of Food Safety and Food Quality, Laboratory of Food Preservation and Food Microbiology, Faculty of Bioscience Engineering, Ghent University, Coupure Links, 653, 9000 Ghent, Belgium.

References

Abreu IMO, Junqueira AMR, Peixoto JR, Oliveira SA (2010) Qualidade microbiológica e produtividade de alface sob adubação química e orgânica. Ciênc Tec Alim 30:108–18

Allende A, Selma MV, Lopez-Galvez F, Villaescusa R, Gil MI (2008) Impact of wash water quality on sensory and microbial quality, including Escherichia coli cross-contamination, of fresh-cut escarole. J Food Prot 71:2514–8

American Public Health Association-APHA (1998) Standard Methods for the Examination of Water and Wastewater. 20 th ed. APHA, Washington, 10 cap

Anderson SA, Turner SJ, Lewis GD (1997) Enterococci in the New Zealand environment: implications for water quality monitoring. Water Sci Tech 35:325–31, 10.1016/S0273-1223(97) 00280-1

Antunes MA (2009) Contaminação, crescimento e inativação de microrganismos na cadeia de produção de alface (Lactuca sativa L.) variedade Vitória de Santo. Dissertation, Federal University of Viçosa

AOAC International (1998) Official methods of analysis of AOAC international (20th ed.) Gaithersburg

Aruscavage D, Lee K, Miller S, LeJeune JT (2006) Interactions affecting the proliferation and control of human pathogens on edible plants. J Food Sci 71(8):R89–99, 10.1111/j.1750-3841.2006.00157

Beuchat LR (2006) Vectors and condition for pre-harvest contamination of fruits and vegetables with pathogens capable of causing enteric diseases. British Food J 108:38–53

Beuchat LR (1996) Pathogenic microorganisms associated with fresh produce. J Food Prot 59:204–16

Buchholz U, Bernard H, Werber D, Böhmer MM, Remschmidt C, Wilking H, Deleré Y, An der Heiden M, Adlhoch C, Dreesman J, Ehlers J, Ethelberg S, Faber M, Frank C, Fricke G, Greiner M, Höhle M, Ivarsson S, Jark U, Kirchner M, Koch J, Krause G, Luber P, Rosner B, Stark K, Kühne M (2011) German outbreak of Escherichia coli O104:H4 associated with sprouts. N Engl J Med 365(19):1763–70, doi:10.1056/NEJMoa1106482

Brasil (2001) Agência Nacional de Vigilância Sanitária (ANVISA) (2001) Resolução RDC n 12, de 02 de janeiro de 2001. Regulamento Técnico sobre padrões microbiológicos para alimentos

Brasil (2013) Empresa Brasileira de Pesquisa Agropecuária - EMBRAPA (2013) Desempenho produtivo de cultivares de alface crespa. Boletim de Pesquisa e Desenvolvimento. ISSN 1677 – 2229

Buck JW, Walcott RR, Beuchat LR (2003) Recent trends in microbiological safety of fruits and vegetables. Plant Health Progress doi:10.1094/PHP-2003-0121-01-RV

Callejón RM, Rodríguez-Naranjo MI, Ubeda C, Hornedo-Ortega R, Garcia-Parrilla MC, Troncoso AM (2015) Reported foodborne outbreaks due to fresh produce in the United States and European Union: trends and causes. Food Path Dis 12(1):32–8, 10.1089/fpd.2014.1821

CDC - Codex Alimentarius Commission (2003) Code of Hygienic Practice for Fresh Fruits and Vegetables Food and Agricultural Organization, Rome. Available via http://www.fao.org/ag/agn/CDfruits_en/others/docs/alinorm03a.pdf

Castro-Ibañez I, Gil MI, Allende A (2013) Impact of extreme climatic events on microbial safety of leafy greens: flooding. Paper presented at the IAFP Annual Meeting, Charlotte, North Carolina

Cevallos-Cevallos JM, Danyluk MD, Gu GY, Vallad GE, van Bruggen AHC (2012) Dispersal of Salmonella Typhimurium by rain splash onto tomato plants. J Food Prot 75:472–9

CONAMA (2005) Resolução 357 de 17 de março de 2005. Available via http://www.mma.gov.br/port/conama/legiabre.cfm?codlegi=459

Delaquis PS, Bach LD, Dinu LS (2007) Behavior of Escherichia coli O157:H7 in leafy vegetables. J Food Prot 70:1966–1974

EFSA - European Food Safety Authority (2014). Scientific Opinion on the risk posed by pathogens in food of non-animal origin. Part 2 (Salmonella and Norovirus in leafy greens eaten raw as salads). EFSA J:12 (3)

FAOSTAT (2013). Food and Agriculture Organization Corporate Statistical Database. In 541 http://faostat3.fao.org/home/index.html#HOME.

FDA - The Food and Drug Administration (1998) Guide to Minimize Microbial food safety hazards for fresh fruits and vegetables. Available via http://www.fda.gov/downloads/Food/GuidanceComplianceRegulatoryInformation/GuidanceDocuments/ProduceandPlanProducts/UCM169112.pdf

Fischer-Arndt M, Neuhoff D, Tamm L, Köpke U (2010) Effects of weed management practices on enteric pathogen transfer into lettuce (Lactuca sativa var. capitata). Food Contr 21(7):1004–10

Franz E, van Diepeningen AD, de Vos OJ, van Bruggen AH (2005) Effects of cattle feeding regimen and soil management type on the fate of Escherichia coli O157:H7 and Salmonella enterica serovar Typhimurium in manure, manure-amended soil, and lettuce. Appl Environ Microbiol 71:6165–74

Girardin H, Morris CE, Albagnac C, Dreux N, Glaux C, Nguyen-The C (2005) Behaviour of the pathogen surrogates Listeria innocua and Clostridium sporogenes during production of parsley in fields fertilized with contaminated amendments. Fems Microbiol Ec 54:287–95

Heaton JC, Jones K (2007) Microbial contamination of fruit and vegetables and the behaviour of enteropathogens in the phyllosphere: a review. J Appl Microbiol 104:613–26, doi:10.1111/j.1365-2672.2007.03587.x

Ilic S, Rajic A, Britton C, Grasso E, Wilkens W, Totton S, Wilhelm B, Waddell L, LeJeune J (2012) A scoping study characterizing prevalence, risk factor and intervention research, published between 1990 and 2010, for microbial hazards in leafy green vegetables. Food Contr 23:7–19, doi:10.1016/j.foodcont.2011.06.027

Itohan AM, Peters O, Kolo I (2011) Bacterial contaminants of salad vegetables in Abuja Municipal Area Concil. Nigéria Mal J Microbiol 7(2):111–4

Jacxsens L, Uyttendaele M, Devlieghere F, Rovira J, Oses Gomez S, Luning PA (2010) Food safety performance indicators to benchmark food safety output of food safety management systems. Int J Food Microbiol 141:S180–S187.

James J (2006) Microbial hazard identification in fresh fruits and vegetables. Wiley Interscience, Dublin

Jaffee S, Masakure O (2005) Strategic use of private standards to enhance international competitiveness: vegetable exports from Kenya and elsewhere. Food Pol 30(3):316–33

Javesnik M, Hlebec V, Raspor P (2008) Food safety knowledge and practices among food handlers in Slovenia. Food Cont 19(12):1107–18

Kirezieva K, Lunning PA, Jacxsens L, Allende A, Johanennssen GS, Tondo EC, Rajkovic A, Uyttendaele M, van Boekel MAJS (2015) Facotrs affecting the status of food safety management systems in the global fresh produce chain. Food Contr 52:85–97, doi:10.1016/j.foodcont.2014.12.030

Kirezieva K, Jacxsens L, Uyttendaele M, Van Boekel M, Luning P (2013)a Assessment of food safety management systems in the global fresh produce chain. Food Res Int 52(1):230-242. doi:10.1016/j.foodres.2013.03.023

Kirezieva K, Nanyunja J, Jacxsens L, Uyttendaele M, Van der Vorst J, Luning P (2013)b Context factors affecting design and operation of food safety management systems in the fresh produce chain. Trends Food Sci Tech 32(2):108–27, doi:10.1016/j.tifs.2013.06.001

Levantesi C, Bonadonna L, Briancesco R, Grohmann E, Toze S, Tandoi V (2012) Salmonella in surface and drinking water: occurrence and water-mediated transmission. Food Res Int 45(2):587–602, doi:10.1016/j.foodres.2011.06.037

Liu C, Hofstra N, Franz E (2013) Impacts of climate change on the microbial safety of pre-harvest leafy green vegetables as indicated by Escherichia coli O157 and Salmonella spp. Int J Food Microbiol 163:119–28

Luning PA, Marcelis WJ, van Boekel MAJS, Rovira J, Uyttendaele M, Jacxsens L (2011) A tool to diagnose context riskiness in view of food safety activities and microbiological safety output. Trends Food Sci Tech 22(1):S67–79, doi:10.1016/j.tifs.2010.09.009

MAFF, The Ministry of Agriculture Fisheries and Food (2000) A study of on-farm manure applications to agricultural land and an assessment of the risks of pathogen transfer into the food chain. Avaiable via http://www.safeproduce.eu/Pics/FS2526.pdf

Mattos LM, Moretti CL, Chitarra AB, Prado MET (2007) Qualidade de Alface Crespa Minimamente Processada Armazenada Sob Refrigeração em Dois Sistemas de Embalagem. Hort Bras 25(4):504–8

Millner P (2003) Composting: improving on a time-tested technique. Agric Res 51(8):20–1

Mocelin AFB, Figueiredo PMS (2009) Avaliação microbiológica e parasitológica das alfaces comercializadas em São Luiz – MA. Rev Inv Biom Uniceuma 1:97–107

Morgharbel ADI, Masson, ML (2005) Perigos associados ao consumo da alface, (Lactuca sativa), in natura. Available via http://serv-bib.fcfar.unesp.br/seer/index.php/alimentos/article/viewFile/105/118

Moyne A, Sudarshana MR, Blessington T, Koike ST, Cahn MD, Harris LJ (2011) Fate of Escherichia coli O157:H7 in field-inoculated lettuce. Food Microbiol 28(8). doi:10.1016/j.fm.2011.02.001

Nicholson FA, Groves SJ, Chambers BJ (2005) Pathogen survival during livestock manure storage and following land application. Bio Tech 96(2):135–43, doi:10.1016/j.biortech.2004.02.030

Nieto-Montenegro S, Brown JL, LaBorde LF (2008) Development and assement of pilot food safety educational materials and training strategies for Hispanic workers in the mushroom industry using the Health Action Model. Food Contr 19(6):616–33

Olaimat AN, Holley RA (2012) Factors influencing the microbial safety of fresh produce: a review. J Food Prot 32(1):1–19, doi:10.1016/j.fm.2012.04.016

Oliveira M, Viñas I, Usall J, Anguera M, Abadias M (2012) Presence and survival of Escherichia coli O157:H7 on lettuce leaves and in soil treated with contaminated compost and irrigation water. Int J Food Microbiol 156(2):133–40, doi:10.1016/j.ijfoodmicro.2012.03.014

Osés SM, Luning PA, Jacxsens L, Santillana S, Jaime I, Rovira J (2012) Microbial performance of food safety management systems implemented in the lamb production chain. J Food Prot 75(1):95–103, doi:10.4315/0362-028X. JFP-11-263

Powell DA, Jacob CJ, Chapman BJ (2011) Enhancing food safety culture to reduce rates of foodbone illness. Food Contr 22(6):817–22

Richardson HY, Nichols G, Lane C, Lake IR, Hunter PR (2009) Microbiological surveillance of private water supplies in England: the impact of environmental and climate factors on water quality. Water Res 43(8):2159–68

Rodrigues RQ, Loiko MR, De Paula CMD, Hessel CT, Jacxsens L, Uyttendaele M, Bender RJ, Tondo EC (2014) Microbiological contamination linked to implementation of good agricultural practices in the production of organic lettuce in Southern Brazil. Food Contr 42:152–64, doi:10.1016/j.foodcont.2014.01.043

Rose JB, Epstein PR, Lipp EK, Sherman BH, Bernard SM, Patz JA (2001) Climate variability and change in the United States: potential impacts on water- and foodborne diseases caused by microbiologic agents. Environ Health Perspct 109(2):211–21

Sala FC, Costa CP (2012) Retrospectiva e tendência da alfacicultura brasileira. Hort Bras 30:187–94

Salem IB, Ouardani I, Hassine M, Aouni M (2011) Bacteriological and physico-chemical assessment of wastewater in different region of Tunisia: impact on human health. BMC Res Notes 4(144). doi:10.1016/S0168-1605(00)00288-9

Sivapalasingam S, Friedman CR, Cohen L, Tauxe RV (2004) Fresh produce: a growing cause of outbreaks of foodborne illness in the United States, 1973 through 1997. J Food Prot 67:2342–53

Stine SW, Song I, Choi CY, Gerba CP (2005) Application of microbial risk assessment to the development of standards for enteric pathogens in water used to irrigate fresh produce. J Food Prot 68:913–8

Taban BM, Halkman AK (2011) Do leafy green vegetables and their ready-to-eat (RTE) salads carry a risk of foodborne pathogens? Anaerobe 17(6):286–7, doi:10.1016/j.anaerobe.2011.04.004

Tirado MC, Clarke R, Jaykus LA, McQuatters-Gollop A, Frank JM (2010) Climate change and food safety: a review. Food Res Int 43:1745–65

Uyttendaele M, Moneim AA, Ceuppens S, El Tahan F (2014) Microbiological safety of strawberries and lettuce for domestic consumption in Egypt. j Food Process. Technol 5:1–7, doi:10.4172/2157-7110.1000308

Warriner K, Huber A, Namvar A, Fan W, Dunfield K (2009) Recent advances in the microbial safety of fresh fruits and vegetables. In: Taylor SL (ed) Adv Food Nut, vol 57., pp 155–208

WHO, World Health Organization, Food and Agriculture Organization of the United Nations (2008) Microbiological risk assessment series: Microbiological hazards in fresh fruits and vegetables. Avaiable via http://www.fao.org/ag/agn/agns/files/FFV_2007_Final.pdf

6

Pesticide contaminants in *Clarias gariepinus* and *Tilapia zilli* from three rivers in Edo State, Nigeria; implications for human exposure

Lawrence I Ezemonye[1], Ozekeke S Ogbeide[1*], Isioma Tongo[1], Alex A Enuneku[1] and Emmanuel Ogbomida[2]

Abstract

The concentrations of 16 pesticides residues (alpha BHC, gamma BHC (lindane), beta BHC, Heptachlor, Aldrin, Heptachlor Epoxide, Endosulfan I, Dieldrin, Endrin, Endosulfan II, 4, 4 DDT, Endosulfan Aldehyde, Endosulfan sulfate, Atrazine, Phosphomethylglycine and Carbofuran) in two fish species (*Tilapia zilli* and *Clarias gariepinus*) from selected rivers (Illushi, Owan and Ogbesse) in Edo State, Nigeria was investigated and the associated human health risks from the consumption of contaminated fishes was also determined. Fish species were collected for a period of 18 months, and pesticides analyses was carried out using Gas Chromatography (GC) equipped with Electron Capture Detector (GC-ECD). The distribution of pesticide residues was more in the tissues of *Clarias gariepinus* (5.53-9.98 µg/g wet weight) than *Tilapia zilli* (3.49- 4.98 µg/g wet weight), while the most dominant pesticide in all the stations and all fish species was the persistent organochlorine; ∑BHC. Furthermore, the estimated dose for alpha BHC, beta BHC, Endosulfan Aldehyde, DDT, Endosulfan I, Endosulfan II, gamma BHC, Heptachlor, Endosulfan sulfate, Atrazine, Phosphomethylglycine and Carbofuran, do not pose direct hazard to human health since values were lower than the reference dose and Hazard quotient (HQ) were lower than toxic threshold of 1. However, estimated dose for, Heptachlor epoxide, Dieldrin, Endrin and Aldrin exceeded the reference dose and were higher than toxic threshold indicating a potential toxicity in humans.

Keywords: Pesticides; Clarias gariepinus; Tilapia zilli; Dietary intake; Health risk; Exposure assessment

Background

The need to produce a greater quantity and quality of food by pest control has resulted in the intensive use of pesticides (Chaverri et al., 2000). This has led to tremendous benefits in the area of agriculture, forestry, public health, and domestic sphere and has also resulted in an economic boom. This scenario has made pesticide an indispensable tool in agricultural production to the extent, one-third of the agricultural products are produced using pesticides of various sorts (Liu et al., 2002). Despite these overwhelming credits of pesticides use, there has been serious health implications to man and his environment accrued from the use of pesticides. These health implications range from the potential risks to human health from both occupational and non-occupational exposures, the death of farm animals and alteration of the local environment (Hossain et al. 2013). Other effects include; immunologic, teratogenic, carcinogenic, reproductive and neurological problems (Babu et al., 2005). Because of these health implications, the use of most classes of pesticides have been banned in developed and some developing countries (including Nigeria) especially the organochlorines (Ize-Iyamu et al., 2007). Inspite of this ban, pesticides especially organochlorines are still major pollutants in Nigerian waters because of the weak enforcement program on the usage of pesticides (Olatunbosun et al., 2011; Upadhi and Wokoma, 2012; Williams, 2013; Ezemonye et al. 2008a, Ezemonye et al., 2008b, Ezemonye et al., 2009; Ize-Iyamu et al., 2007; Adeboyejo et al., 2011; Okeniyia et al., 2009; Adeyemi et al., 2011).

* Correspondence: oze_ogbe@yahoo.com
[1]Ecotoxicology and Environmental Forensics Laboratory, University of Benin, Benin City, Edo State, Nigeria
Full list of author information is available at the end of the article

Food consumption has been identified as an important route of human exposure to pesticides with concentrations of pesticides in fish leading to several health concerns, particularly for high-risk population groups, such as pregnant women and children (USEPA, 1998; Jaing et al. 2005). Although the presence of trace levels of these pesticides in food is considered as an indication that contamination has occurred, the risk of adverse health effects also depends on their concentration, frequency of contact and duration of exposure (Jiang et al., 2005). Therefore, assessing the health risk of pesticide exposure involves the use ecological risk assessment models (ERAs). ERA is a process of assessing the ecological health of species, populations, communities and ecosystems due to pesticide exposure (Hoffman et al., 2003). This assessment is necessary because of the tendency of pesticides to accumulate and persist in body tissues, leading to acute or chronic health effects (Pardío et al., 2012).

Assessing Human Health risk involves the use of food consumption data that have been established for a country (Jiang et al., 2005), body weight estimates and acceptable daily intake (ADI) or reference dose set by various regulatory bodies. FAO defines "ADI" of a chemical as the daily intake which, during an entire lifetime, appears to be without appreciable risk to the health of the consumer (WHO, 1997). In recent times, researchers have begun to estimate the risk posed to humans via the consumption of contaminated foods stuffs. Several studies including; Jiang et al. (2005), Gavor et al. (2013), Yohannes et al. (2013), Darko and Akoto (2007), Andoh et al. (2013), Fianko et al. (2011) have reported potential human health risk from the consumption of contaminated food. In Nigeria, there is the dearth of data on the concentration of pesticide residues in various water bodies, while information on the human health effects associated with consumption of aquatic organisms from the polluted water bodies is nonexistent to the best of the author's knowledge (Ize-Iyamu et al., 2007; Ezemonye et al., 2008a,b and Okoya et al., 2013).

This study was conducted in order to assess the potential health risks associated with the consumption of two common fish species contaminated with pesticide residues, obtained from three rivers in Edo State.

Methods
Study area and sample collection
The rivers selected were Illushi River (N: 06° 45' 40"; E: 005° 46' 07.4"), Owan River (N: 06° 39' 59.8"; E: 006° 36' 34.2") and Ogbesse river (N: 06° 45' 3.7"; E: 005° 34' 03.2") (Figure 1). A total of 54 samples of *Clarias gariepinus* and *Tilapia zilli* each were simultaneously collected for this study. Sampling period lasted for 18 months (January 2012 to June 2013).

Extraction of pesticide residues in fish samples
A simple, rapid solvent extraction method was used to determine total lipids in each fish tissue based on the method described by Randall et al. 1998 and lipid normalized concentrations were obtained using the ratio between pesticide concentration in tissue and lipid fraction in the tissue.

The frozen composite edible portions of tissue samples for each species (*Clarias gariepinus* and *Tilapia zilli*) were used for extraction based on the method described by Steinwandter (1992). 25 g of the samples was inserted into a homogenizer cup, and 100 ml of acetone was added. The sample was homogenized for 20 minutes at 100 rpm. The sample was further mixed with 5 g of anhydrous sodium sulphate. Extraction was done using a soxhlet extraction for approximately 20–25 minutes using dichloromethane and n hexane mixture. The resulting extract was dissolved with hexane and re-concentrated to 1 to 3 ml (USEPA, 2004).

Gas chromatographic (GC) analysis
The cleaned up extracts were analyzed for pesticides (α BHC, γ - BHC, β- BHC, Heptachlor, Heptachlor Epoxide, Aldrin, Dieldrin, Endrin, 4,4 DDT, Endosulfan I, Endosulfan II, Endosulfan aldehyde, Endosulfan sulfate, Atrazine, Phosphomethylglycine and Carbofuran). Results were obtained using a Hewlett-Packard (*hp*) 5890 Series II equipped with 63Ni Electron Capture Detector (ECD) of activity 15 mCi with an auto sampler. The chromatographic separation was done using a VF-5 ms of 30 mm capillary column with 0.25 mm internal diameter and 0.25 μm film thicknesses and equipped with 1 m retention gap (0.53 mm, deactivated). The GC conditions were as follows: The oven temperature programme: Initial temperature was set at 60°C for 2 min and ramped at 25°C/min to 300°C for 5mins and allowed to stay for 15 min giving a total run time of 58 min. The injector setting was a pulsed spit less mode with a temperature of 250°C at a standard pressure. The injection volume was 1.5 ml. The detector temperature was 320°C (held for 5 minutes), Helium was used as a carrier gas while Nitrogen gas (N_2) was used as the makeup gas, maintained at a constant flow rate of 29 ml/min.

The efficiency of the analytical method (the extraction and clean-up methods) was determined by recoveries of an internal standard. Peak identifications were conducted by comparing the retention time of standards and those obtained from the extracts. Concentrations were calculated using a four-point calibration curve. Method detection limits (MDLs) ranged from 0.01 μg/g/ dw for pesticides in biota.

Health risk estimation
To assess the risk of pesticide contained in each fish species on consumers, the guide- lines for potential risk

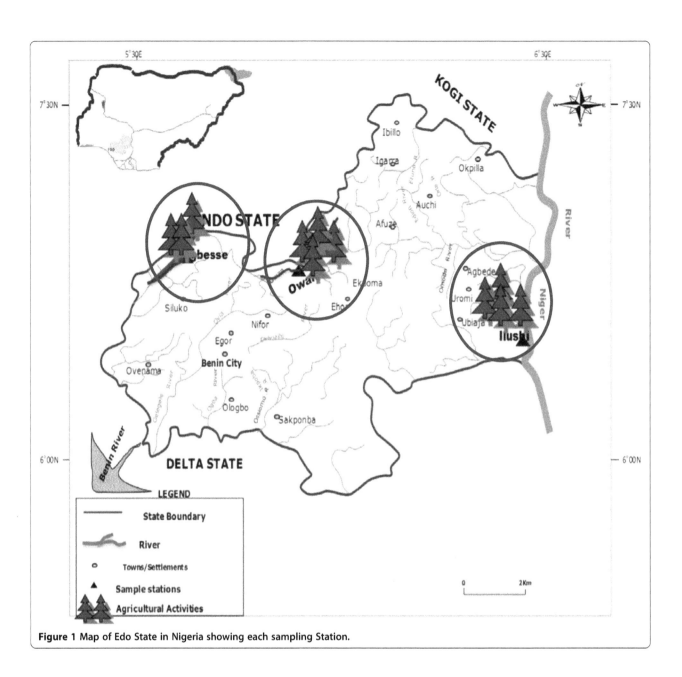

Figure 1 Map of Edo State in Nigeria showing each sampling Station.

assessment drawn up by the US EPA were used. The reference dose (RfD) of each pesticides is the exposure that is likely to be without an appreciable risk of deleterious effects and was provided by the US EPA (USEPA, 2006). Based on estimated average daily intake (EADI, with Eq. (1)), hazard quotients (HQ, with Eq. (2)) were calculated to estimate the non-cancer risks of pesticide exposure.

Estimated average daily intake (EADI)

The estimated Average Daily intake (EADI) was found by multiplying the average/ mean residual pesticide concentration (μg/ww) by the fish consumption rate (Kg/day).

$$\text{EADI } (\mu g/kg/day) = \frac{\text{Residual concentration } \times \text{ Food consumption/}}{\text{Body weight.}}$$

$$(1)$$

(WHO, 1997; Fianko et al., 2011),

Hazard quotient (HQ)

Hazards quotients were obtained by dividing the EADI by their corresponding reference dose (RfD)

$$\text{Hazard Quotient} = \frac{\text{Estimated Average Daily Intake}}{\text{Reference Dose}}$$

$$(2)$$

(WHO, 1997; Fianko et al., 2011),

The food and agricultural organization (FAO 2011) quotes the per capita consumption of fishes in Nigeria as 9 kg. While body weight was set at 70 kg for adult population group.

Results

Analyses showed levels of pesticides residues in 40 samples of *Clarias gariepinus* and 45 samples of *Tilapia zilli* (Illushi River), 49 samples of *Clarias gariepinus* and 48 samples of *zilli* (Owan River) in Edo State. These findings are summarized in Tables 1 and 2 respectively. The concentration profile of pesticides in each fish species is also presented in Figure 2. It was observed that all pesticides residues detected showed higher concentrations in *Clarias gariepinus* than *Tilapia zilli* with beta BHC having the highest concentration 1.67 & 0.93 µg/g ww respectively. In all the rivers sampled, total pesticide residues was observed to be more in the tissues of *Clarias gariepinus* (5.53-9.98 µg/g ww) than *Tilapia zilli* (3.49- 4.98 µg/g ww). However there was no significant difference between concentrations of pesticide residues in Clarias gariepinus and Tilapia zilli (p> 0.05; F = 3.23) in each of the river. Spatial variation showed that Ogbesse River had the highest concentration of pesticide residues in both fish species when compared with Owan and Illushi Rivers (Figure 3). The Organochlorine pesticide beta BHC was the most dominant pesticide, having the highest mean concentration (0.93 µg/g ww and 1.67 µg/g ww) and percentage distribution (49% each) in *Clarias gariepinus* and *Tilapia zilli* respectively (Figure 4 and 5).

Dietary intake levels

The Estimated Acceptable Daily Intake (EADI) of pesticides in *Clarias gariepinus* and *Tilapia zilli* for adult population in Edo State is presented in Table 3. Consumptions rates of fishes in Nigeria were obtained from FAO statistics book (FAO, 2011). Risk estimations in *Tilapia zilli* showed that Heptachlor epoxide and Aldrin exceeded the reference dose and toxicity threshold level (1). While in *Clarias gariepinus*, Heptachlor epoxide, Aldrin, Dieldrin and Endrin exceeded the reference dose and toxicity threshold level (1) (Figure 6).

Discussion

The presence of pesticide residues in *Clarias gariepinus* and *Tilapia zilli* obtained from Illushi, Owan and Ogbesse rivers is an evidence of bioconcentration (from water via gills and epithelial tissues) and bioaccumilation (through water and food, leading to biomagnification in different organisms) of pesticides from the surrounding environment (Murty 1986). The levels and occurrence of residues in fish samples seem to be governed by feeding mode, age and mobility of the biota, consequently, higher concentrations of pesticide residues oberved in *Clarias gariepinus* may be attributed to the feeding mode of the fish (Mwevura et al., 2002). Furthermore, Biego et al. (2010), states that *Claris gariepinus* habitates the level of sediments where it gets most of its food, hence the likely hood of exposure to pesticides. Apart from the entry of pesticides through bioconcentration (via water) in to *Clarias gariepinus*, they are bottom

Table 1 Mean concentration (ug/gww) of pesticide residues in *Clarias gariepinus*

Pesticides	ILLUSHI Mean ± Std. D.		Range		OGBESSE Mean ± Std. D.		Range		OWAN Mean ± Std. D.		Range	
alpha – BHC	1.48	±2.71	0	8.90	1.86	±1.85	0	7.30	1.16	±2.05	0	6.50
gamma-BHC	0.68	±1.35	0	5.60	0.66	±0.64	0	2.10	0.66	±1.10	0	3.70
beta – BHC	1.58	±2.48	0	6.40	2.34	±2.89	0	8.30	1.09	±1.20	0	4.30
Heptachlor	0.20	±0.37	0	1.50	0.43	±0.79	0	3.00	0.17	±0.46	0	1.90
Heptachlor epoxide	0.21	±0.48	0	1.80	0.25	±0.43	0	1.10	0.39	±0.62	0	2.00
Aldrin	0.82	±1.62	0	6.50	0.63	±0.80	0	2.15	0.12	±0.24	0	0.80
Dieldrin	0.44	±1.57	0	6.70	0.53	±0.90	0	2.50	0.38	±1.10	0	4.70
Endrin	0.31	±0.53	0	1.90	0.45	±1.17	0	4.90	0.39	±0.64	0	2.20
DDT	0.03	±0.10	0	0.40	0.04	±0.15	0	0.60	0.09	±0.20	0	0.70
Endosulfan I	0.25	±0.57	0	2.40	0.13	±0.40	0	1.70	0.39	±0.96	0	4.10
Endosulfan 11	0.11	±0.25	0	0.80	0.38	±1.25	0	5.30	0.05	±0.21	0	0.90
Endosulfan aldehyde	0.38	±1.24	0	5.30	0.27	±0.52	0	1.40	0.06	±0.19	0	0.70
Endosulfan sulfate	0.42	±1.72	0	7.30	0.26	±0.99	0	4.20	0.02	±0.07	0	0.30
Phosphomethylglycine	0.31	±0.74	0	3.10	0.91	±2.18	0	7.20	0.10	±0.18	0	0.60
Atrazine	0.11	±0.21	0	0.60	0.73	±1.22	0	3.50	0.13	±0.45	0	1.90
Carbofuran	0.78	±2.13	0	7.80	0.22	±0.92	0	3.90	0.32	±0.57	0	2.10

Table 2 Mean concentration (ug/gww) of pesticide residues in *Tilapia zilli*

Pesticides	ILLUSHI				OGBESSE				OWAN			
	Mean ± Std. D.		Range		Mean ± Std. D.		Range		Mean ± Std. D.		Range	
alpha – BHC	0.38	±0.90	0	3.80	0.88	±1.36	0	4.00	0.59	±1.41	0	5.50
gamma - BHC	0.85	±2.04	0	8.80	0.13	±0.21	0	0.70	0.34	±0.68	0	2.40
beta – BHC	0.29	±0.35	0	1.00	1.73	±2.57	0	8.00	0.77	±1.69	0	6.00
Heptachlor	0.60	±0.98	0	2.90	0.74	±1.51	0	6.00	0.22	±0.46	0	1.90
Heptachlor epoxide	0.12	±0.27	0	1.00	0.17	±0.25	0	0.60	0.06	±0.13	0	0.50
Aldrin	0.73	±1.64	0	5.20	0.12	±0.24	0	0.70	0.13	±0.16	0	0.50
Dieldrin	0.08	±0.18	0	0.60	0.08	±0.15	0	0.50	0.17	±0.24	0	0.80
Endrin	0.02	±0.07	0	0.30	0.07	±0.16	0	0.60	0.06	±0.11	0	0.40
DDT	0.00	±0.00	0	0.00	0.08	±0.23	0	0.80	0.03	±0.12	0	0.50
Endosulfan I	0.17	±0.34	0	1.40	0.22	±0.32	0	1.10	0.19	±0.28	0	1.00
Endosulfan 11	0.03	±0.10	0	0.30	0.03	±0.14	0	0.60	0.02	±0.07	0	0.30
Endosulfan aldehyde	0.12	±0.29	0	0.90	0.02	±0.06	0	0.20	0.16	±0.34	0	1.20
Endosulfan sulfate	0.03	±0.12	0	0.50	0.16	±0.43	0	1.70	0.09	±0.29	0	1.20
Phosphomethylglycine	0.15	±0.25	0	0.90	0.42	±0.89	0	3.00	0.54	±1.25	0	5.00
Atrazine	0.18	±0.29	0	1.00	0.09	±0.14	0	0.40	0.09	±0.21	0	0.60
Carbofuran	0.09	±0.24	0	0.90	0.02	0.09	0	0.40	0.03	±0.08	0	0.30

feeders hence they are in close proximity to contaminated sediments. Mwevura et al. (2002) reported that biota in close proximity to sediments pick up residues from the sediment and this occurs by passive equilibration process through their membranes. Murano et al. (1997) and Kidwell et al. (1990) equally adds that pesticide accumulation in fish was due to their lipid content, this implies that the high lipid content in *Clarias gariepinus*, allows more pesticide residues tend to be trapped in their lipid stores compared to Tilapia zilli. Romanic et al., (2014) have observed a positive correlation between the lipd content of fish muscles and the concentration of organochlorine pesticides.

This study also corroborates with Deribe et al., (2014) who reports that variation in the accumulation of POPs in different fish species is attributed to trophic position, age, and fat content. The levels of pesticide contamination in fish species observerd in this study were higher

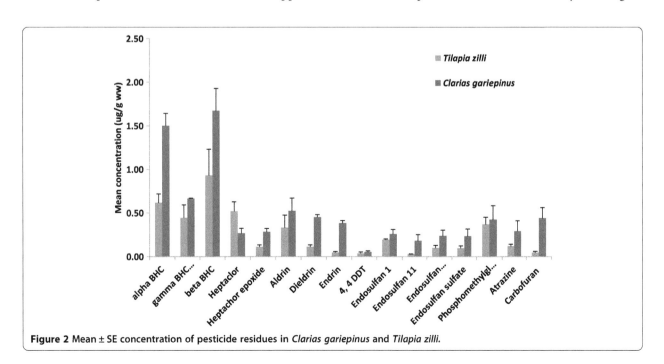

Figure 2 Mean ± SE concentration of pesticide residues in *Clarias gariepinus* and *Tilapia zilli*.

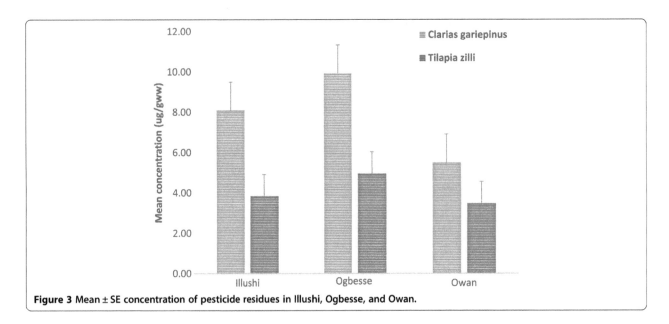

Figure 3 Mean ± SE concentration of pesticide residues in Illushi, Ogbesse, and Owan.

than concentrations obtained in earlier studies in Nigeria. Some of these studies include: Upadhi and Wokoma (2012), Unyimadu and Udochu, (2002); Ize-Iyamu et al., (2007); Williams (2013). However concentrations observed in this study was lower than the concentrations reoprted by Ezemonye et al., (2009); Adeboyejo et al., (2011).

The presence of pesticides in fishes is a major concern because pesticides have a number of adverse effects on the aquatic organism such as reproductive impairment and suppression of the immune system (Aguilar et al., 2002), which can have long-term consequences for population viability. Furthermore, consumption of each fish species, especially species with more fat content (*Clarias gariepinus*), and from high trophic levels, may expose consumers to possible health hazard because the consumption of contaminated food (including fish) has been established as a major route of human exposure to pesticides and other contaminants (Biego et al., 2010; Ni et al., 2012; Barnhoorn et al., 2015).

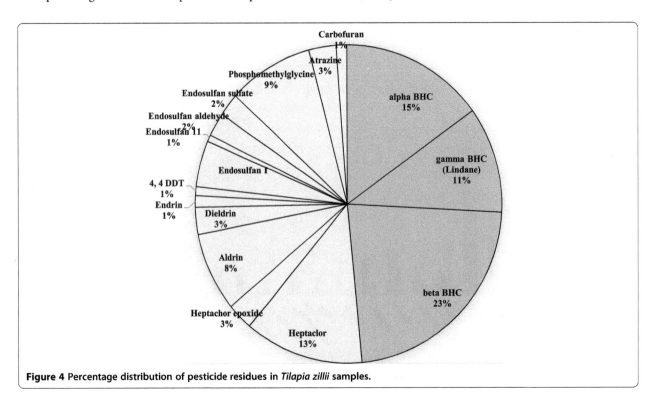

Figure 4 Percentage distribution of pesticide residues in *Tilapia zillii* samples.

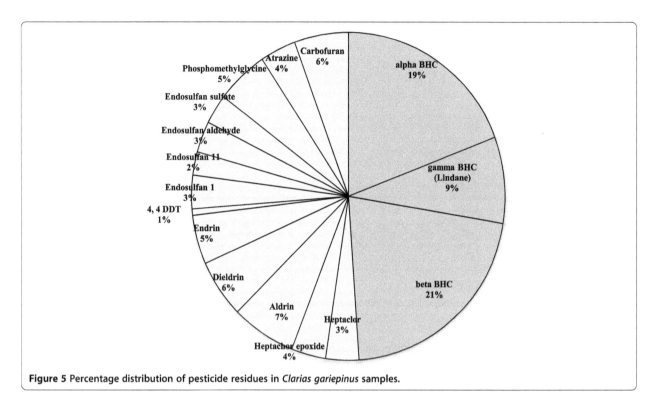

Figure 5 Percentage distribution of pesticide residues in *Clarias gariepinus* samples.

Health risk estimations

Assessment of the human health risk to estimate the negative and non-carcinogenic impact associated with the consumption of *Clarias gariepinus* and *Tilapia zilli*, reveals potential hazard to humans. *Clarias gariepinus* and *Tilapia zilli* are commercial aquatic products in Edo state and are a readily available and cheap source of protein for most families. Hence assessing the potential human health risk from the consumption of these fishes is an extremely important step towards public health safety. EADI for Heptachlor and Aldrin in *Tilapia zilli* were higher than their recommended reference dose

Table 3 Health risk assessment of pesticide residues in fishes from rivers in Edo State

Pesticides	Rfd (µg/kg/day)	Tilapia zilli			Clarias gariepinus		
		Residual conc	EADI	HQ	Residual conc	EADI	HQ
alpha – BHC	8	0.62	0.08	0.01	1.50	0.19	0.02
gamma - BHC	0.3	0.44	0.06	0.19	0.67	0.09	0.29
beta – BHC	8	0.93	0.12	0.01	1.67	0.21	0.03
Heptachlor	5	0.52	0.07	0.01	0.27	0.03	0.01
Heptachlor epoxide	**0.013**	0.11	**0.01**	**1.09**	0.28	**0.04**	**2.77**
Aldrin	**0.03**	0.33	**0.04**	**1.41**	0.52	**0.07**	**2.23**
Dieldrin	0.05	0.11	0.01	0.28	0.45	**0.06**	**1.16**
Endrin	0.03	0.05	0.01	0.21	0.38	**0.05**	**1.63**
DDT	0.5	0.04	0.01	0.01	0.05	0.01	0.01
Endosulfan 1	6	0.19	0.02	0.00	0.26	0.03	0.01
Endosulfan 11	6	0.03	0.00	0.00	0.18	0.02	0.00
Endosulfan aldehyde	6	0.10	0.01	0.00	0.24	0.03	0.01
Endosulfan sulfate	6	0.10	0.01	0.00	0.23	0.03	0.00
Phosphomethylglycine	300	0.37	0.05	0.00	0.42	0.05	0.00
Atrazine	35	0.12	0.02	0.00	0.29	0.04	0.00
Carbofuran	5	0.05	0.01	0.00	0.44	0.06	0.01

Rfd = (Reference Dose). HQ = Hazard quotient, EADI = Estimated Average Daily Intake. Bold figures are above the thresh hold limit.

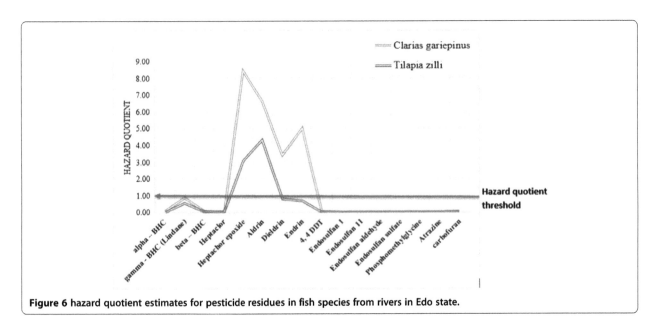

Figure 6 hazard quotient estimates for pesticide residues in fish species from rivers in Edo state.

while Hazard Quotient (HQ) values were above one (1) suggesting that the potential health risks associated with intake of Heptachlor and Aldrin, through the consumption of *Tilapia zilli* was high. On the other hand, EADI for Heptachlor epoxide, Endrin, Aldrin and Dieldrin in *Clarias gariepinus* were higher than their recommended reference dose while estimated HQ were greater than 1 suggesting potential health risk through the consumption of *Clarias gariepinus*. For *Clarias gariepinus* and *Tilapia zilli* the hazard index estimations showed that mixtures of organochlorine pesticides (3.863 and 9.592 respectively) were above one (1) for adult population group, suggesting that toxicity to humans from exposure could be significant. However, hazard estimates for Organophosphate, Triazine and Carbamate pesticides were below one. It has been reported that if the hazard index >1, the mixture has exceeded the maximum acceptable level (Tsakiris et al. 2011) and there might thus be risk. Therefore, results from this study suggest a great potential for chronic toxicity through the consumption of pesticide contaminated. The results of Human Health Risk estimations obtained from this study, conforms with studies by Darko and Akoto (2008), on pesticides in eggplant and tomatoes from Kumasi market, Sohair et al. (2013) on the Estimated Daly Intake and corresponding Health hazard of pesticide residues present in fruits from Egypt, Fianko et al. (2011) on pesticides in fish species from Densu River basin in Ghana and Andoh et al. (2013) pesticides in maize and cowpea from Ghana.

Conclusion

The concentrations of 16 pesticides residues (α BHC, γ - BHC (lindane), β- BHC, Heptachlor, Aldrin, Heptachlor Epoxide, Endosulfan I, Dieldrin, Endrin, Endosulfan II, 4, 4 DDT, Endosulfan Aldehyde, Endosulfan sulfate, Atrazine, Phosphomethylglycine and Carbamate) in two common fish species in Edo State, Nigeria have been determined. From the results obtained, *Clarias gariepinus* had significantly higher concentrations of pesticide residues compared with *Tilapia zilli*. Levels of pesticide observed in both *Clarias gariepinus* and *Tilapia zilli* were lower than the recommended Reference dose except, Heptachlor epoxide, Dieldrin, Endrin and Aldrin. Risk assessment showed that Hazard quotients for Heptachlor epoxide, Dieldrin, Endrin and Aldrin were above the toxic threshold of 1. These results indicate that human exposure (adults) to pesticides via consumption of contaminated fishes could lead to potential health risk. It also substantiate the fact that the long term accumulation of pesticide residues in the human body via dietary intake of fish is a source of concern.

This calls for more stringent regulations on the use of pesticides by farmers and also a rigorous and frequent monitoring of pesticides in various environmental matrix especially food should be encouraged so as to safe guard the health of consumers.

Competing interests
The authors declare that they have no competing interests.

Authors' contributions
L.E was the team leader of this project. He was responsible for providing the frame work and design for the overall project. O.O, was responsible for sample collection, gas chromatography instrumentation and compilation of this paper. I.T was involved in sampling for fishes, extraction and preparation of samples for further analysis, A.E was involved in sampling and the estimations of health risk and statistical analysis. E.O was also responsible for lipid normalization of samples prior to analysis. All authors read and approved the final manuscript.

Author details

[1]Ecotoxicology and Environmental Forensics Laboratory, University of Benin, Benin City, Edo State, Nigeria. [2]National Centre for energy and environment (NCEE), University of Benin, Benin City, Edo State, Nigeria.

References

Adeboyejo OA, Clarke EO, Olarinmoye MO (2011) Organochlorine pesticides residues in water, sediments, fin and shell-fish samples from Lagos lagoon complex, Nigeria. Researcher 3:38–45

Adeyemi D, Anyakora C, Ukpo G, Adedayo A, Darko G (2011) Evaluation of the levels of organochlorine pesticide residues in water samples of Lagos Lagoon using solid phase extraction method. Journal of Environmental Chemistry and Ecotoxicology 3:160–166

Aguilar A, Borrell A, Reijnders PJH (2002) Geographical and temporal variation in levels of organochlorine contaminants in marine mammals. Mar Environ Res 53:25–52

Andoh H, Akoto O, Darko G (2013) Health Risk Assessement of Pesticide residues in maize and cowpea from Ejura, Ghana. Paper presented at 5th International Toxicology Symposium in Africa., Jointly Hosted by College of Science, Kwame Nkrumah University of Science and Technology, Ghana and Hokkaido University, Japan sponsored by Japan Society for Promotion of Science

Babu RR, Imagawa T, Tao H, Ramesh R (2005) Distribution of PCBs, HCHs and DDTs, and their ecotoxicological implications in Bay of Bengal. India Environment International 31:503–512

Barnhoorn IEJ, van Dyk JC, Genthe B, Harding WR, Wagenaar GM, Bornman MS (2015) Organochlorine pesticide levels in Clarias gariepinus from polluted freshwater impoundments in South Africa and associated human health risks. Chemosphere 120(2015):391–397

Biego GHM, Yao KD, Ezoua P, Kouadio LP (2010) Assessment of Organochlorine Pesticides Residues in Fish Sold in Abidjan Markets and Fishing Sites. African Journal of food agriculture Nutrition and development 10(3):2305–2323

Chaverri F, Soto L, Ramirez F, Bravo V (2008) Preliminary diagnosis of pesticide use in crops of rice, bananas, coffee, sugarcane, onion, melon, orange, potato, pineapple, tomato, flowers and ornamentals. Final Project Report Diagnosis pesticide use in Costa Rica. SAREC Programme: Research program in environment and health in Central America. Faculty of Earth Sciences and the Sea Regional Institute for Studies on Toxic Substances (IRET). National University. Heredia, Costa Rica. SAREC Publication: Pg 46

Darko G, Akoto O (2008) Dietary intake of organophosphorus pesticide residues through vegetables from Kumasi, Ghana. Food and Chemical Toxicology 46:3703–3706

Deribe E, Bjørn Olav R, Reidar B, Brit S, Zinabu G, Elias D, Lindis S, Ole Martin E (2014) Organochlorine Pesticides and Polychlorinated Biphenyls in Fish from Lake Awassa in the Ethiopian Rift Valley: Human Health Risks. Bull Environ Contam Toxicol (2014) 93:238–244

Ezemonye LIN, Ikpesu TO, Tongo I (2008a) Distribution of Lindane in Water, Sediment, and Fish from the Warri River of the Niger Delta. Nigeria Journal of Arh Hig Toksikol 59:261–270

Ezemonye LIN, Ikpesu TO, Tongo I (2008b) Distribution of Diazinon in Water, Sediment and Fish from Warri River, Niger Delta, Nigeria. Jordan J Biol Sci 1 (2):77–83

Ezemonye LIN, Ikpesu TO, Tongo I (2009) Distribution of Propoxur in water, sediment and fish from Warri River, Nigeria Delta, Nigeria. Turkish Journal of Biochemistry 34(3):121–127

FAO (2011) Fishery and Aquaculture Statistics. 2009. Statistics and Information Service of the Fisheries and Aquaculture Department/Service. 2009. Rome/ Roma, FAO. 2011., p 78

Fianko RJ, Augustine D, Samuel TL, Paul OY, Eric TG, Theodosia A, Augustine F (2011) Health Risk Associated with Pesticide Contamination of Fish from the Densu River Basin in Ghana. Journal of Environmental Protection 2:115–123

Gavor S, Martin Kwasi A, Osei A (2013) Assessment of Health Risk of Pesticide Residues in Selected Vegetables from Kumasi Central Market. Paper presented at 5th International Toxicology Symposium in Africa. Jointly Hosted by College of Science, Kwame Nkrumah University of Science and Technology. Japan sponsored by Japan Society for Promotion of Science, Ghana and Hokkaido University, Hokkaido University publication

Hoffman DJ, Rattner BA, Burton GA, Cairns J (2003) Handbook of Ecotoxicology, 2nd edn. Lewis Publishers, CRC Press, Boca Raton, FL

Hossain MS, Hossain MA, Rahman MA, Islam MM, Rahman MA, Adyel TM (2013) Health risk assessment of pesticide residues via dietary intake of market vegetables from Dhaka, Bangladesh. Foods 2013(2):64–75

Ize-Iyamu OK, Asia IO, Egwakhide PA (2007) Concentrations of residues from organochlorine pesticide in water and fish from some rivers in Edo State Nigeria. Int J Phys Sci 2(9):237–241

Jiang QT, Lee TKM, Chen K, Wong HL, Zheng JS, Giesy JP, Lo KKW, Yamashita N, Lam PKS (2005) Human health risk assessment of organochlorines associated with fish consumption in a coastal city in China. Environ Pollut 136:155–165

Kidwell JM, Phillips LJ, Birchard GF (1990) Comparative analysis of contaminants levels in bottom feeding and predatory fish using the national contaminants biomonitoring program data. Bull Contam Tox 55(6):919–923

Liu CJ, Men WJ, Liu YJ (2002) The Pollution of Pesticides in Soils and its Bioremediation. System Sciences and Comprehensive Studies in Agriculture 18(4):295–297

Maurano FM, Guida G, Melluso GS (1997) Accumulation of Pesticide Residues in Fishes and Sediments in the River Sele (South Italy). Journal of Preventive Medicine and Hygiene 38:3–4

Murty AS (1986) Toxicity of pesticides to fish Volume II. LRC Press, Boca Roton F. L

Mwevura H, Othman CO, George LM (2002) Organochlorine Pesticide Residues in Edible Biota from the Coastal Area of Dar es Salaam City. Pesticide Residues In Coastal Dar Western Indian Ocean. J Mar Sci 1(1):91–96

Ni H-G, Chao D, Shao-You L, Xiao-Ling Y, Sojinu Olatunbosun S (2012) Food as a main route of adult exposure to PBDEs in Shenzhen, China. Science of the Total Environment 437(2012):10–14

Okeniyia SO, Eqwikhide PA, Akporhonore EE, Obazed EI (2009) Distribution of organochlorine and polychlorinated pesticides residues in water bodies of some rivers in northern Nigeria. EJEAFche 8(11):1269–1274

Okoya AA, Ogunfowokan AO, Asubiojo OI, Torto N (2013) Organochlorine Pesticide Residues in Sediments and Waters from Cocoa Producing Areas of Ondo State. International Scholarly Research Network (ISRN) Soil Science, Southwestern Nigeria

Olatunbosun SS, Sojinu O, Sonibare E, Eddy O, Zeng Y (2011) Occurrence of Organochlorine Pesticides (OCPs) in Surface Sediments of the Niger Delta, Nigeria. Journal of Applied Sciences Research 7(8):1299–1305

Pardío V, Martínez D, Flores A, Romero D, Suárez V, López K, Uscanga R (2012) Human health risk of dietary intake of organochlorine pesticide residues in bovine meat and tissues from Veracruz, México. Food Chemistry 135:1873–1893

Randall RC, Young DR, Lee H II, Echols SF (1998) Lipid methodology and pollutant normalization relationships for neutral nonpolar organic pollutants. Environ Toxicol Chem 17(5):788–791

Romanic SH, Herceg HD, Lazar B, Klincic D, Mackelworth P, Fortuna CM (2014) Organochlorine contaminants in tissues of common bottlenose dolphins Tursiopsruncatus from the northeastern part of the Adriatic Sea. Environmental toxicology and pharmacology 38(2014):469–479

Sohair AG, Mohsen A, Ayoub M, Amer MA, Wasfi MT (2013) Dietary Intake of Pesticide Residues in some Egyptian Fruits. Journal of Applied Sciences Research 9(1):965–973

Steinwandter H (1992) Development of Microextraction methods in residue analysis. In: Cairns T, Sherma J (eds) Emerging strategies for pesticide analysis. CRC, Boca Raton, FL, pp 3–50

Tsakiris Ioannis N, Toutoudaki M, Kokkinakis M, Paraskevi M, Tsatsakis AM (2011) A Risk Assessment Study of Greek Population Dietary Chronic Exposure to Pesticide Residues in Fruits, Vegetables and Olive Oil in Pesticides - Formulations, Effects, Fate. Stoytcheva M (Ed). ISBN: 978- 953-307-532-7

United States Environmental Protection Agency (USEPA) (1998) Guidelines for Ecological Risk Assessment. Washington, US: EPA, Report No. EPA/630/ R-95/002F

United States Environmental Protection Agency (USEPA) (2004) Guidelines for Water Reuse. EPA/625/R-04/108.US EPA. CampDresser & McKee Inc, Washington, DC

United States Environmental Protection Agency (USEPA) (2006) Guidance for assessing Chemical contaminant data foruse in fish advisories, 2, 231-253

Unyimadu JP, Udochu A (2002) Comparative Studies of Organochlorine and PCBs in Fish from the Lagos Lagoon. River Elber Saar. Journal of Agricultural Biotechnology: Environmental 4(1-2):14–17

Upadhi F, Wokoma OAF (2012) Examination of Some Pesticide Residues in Surface Water, Sediment and Fish Tissue of Elechi Creek, Niger Delta. Nigeria Research Journal of Environmental and Earth Sciences 4(11):939–944

WHO (World Health Organization) (1997) Guidelines for predicting dietary intake of pesticide residues (revised) global environmental programme (GEMS/ Food) in collaboration with Codex Committee on pesticide residues. Programme of Food Safety and Food Aid 1–44

Williams BA (2013) Levels and distribution of chlorinated pesticide residues in
    water and sediments of Tarkwa Bay, Lagos Lagoon. Journal of Research in
    Environmental Science and Toxicology 2(1):1–8
Yohannes YB, Ikenaka Y, Nakayama SMM, Ishizuka M (2013) Organochlorines pesticides in
    fish species from Lake Ziway, Ethiopa; Association with Tropic Level and Human
    Health Risk Assessment, Paper presented at 5th International Toxicology Symposium
    in Africa. Jointly Hosted by College of Science, Kwame Nkrumah University of
    Science and Technology, Ghana and Hokkaido University, Japan sponsored by
    Japan Society for Promotion of Science. Hokkaido University publication

7

Fruit juice and puree characteristics influence enrichment requirements for real-time PCR detection of *Alicyclobacillus acidoterrestris*

Shima Shayanfar[1], Christina Harzman[2] and Suresh D Pillai[1*]

Abstract

Background: *Alicyclobacillus acidoterrestris* is a key spoilage causing bacterium commonly found in fruit juices and purees. Commercial real-time PCR based assays to detect this organism are available, but reportedly require 48 hours of enrichment for detection. The underlying hypothesis of this study was that fruit juice and puree characteristics influence the enrichment requirements of this organism, and that in some matrices, the organism can be detected within 24 hours even when present at low initial contamination. Thirteen different store-purchased fruit juice and purees were inoculated with 10 CFU/ml of *Alicyclobacillus acidoterrestris*. The inoculated samples were enriched for 24 and 48 h. Aliquots from the un-enriched, 24 hour, and 48 hour enriched samples were taken, total community DNA extracted, and the real-time PCR assay performed using commercially available kits.

Results: *A. acidoterrestris* was detected by real-time PCR within 24 h of enrichment in most matrices (except ketchup and orange concentrate) even from a low starting concentration (10 CFU/ml). Juice and puree samples with high soluble solids contents (i.e. Brix values) required longer incubation periods for lower *A. acidoterrestris* Ct values.

Conclusions: The soluble solids (Brix) content of fruit juice and purees influence the enrichment requirements for real-time PCR detection of *A. acidoterrestris*. Samples with high Brix values should be diluted to reduce the inhibitors of *A. acidoterrestris* proliferation during enrichment.

Keywords: Alicyclobacillus acidoterrestris; Fruit; Juice; Brix; Real-time PCR; Enrichment; Ct value

Background

Alicyclobacillus acidoterrestris is a thermophilic, acid tolerant, spoilage bacterium that withstands pasteurization temperatures typically employed in the juice industry. As the most commonly detected *Alicyclobacillus* species in spoiled fruit juices and purees, *A. acidoterrestris* converts lignin components to form guaiacol, contributing to the unpleasant 'hammy smell" in fruit juices undergoing spoilage (Eguchi et al. 2000; Silva and Gibbs 2001; Gocmen et al., 2005; Bahçeci et al., 2005; Chang and Kang, 2004; Groenewald et al. 2013).

Alicyclobacillus spp. spoilage is a global problem in the fruit juice industry (Cagnasso et al. 2010; Groenewald et al. 2009, McKnight et al. 2010, Walls & Chuyate 2000). Most instances of contamination have been reported in apple and orange juices (McKnight et al., 2010; Groenewald et al., 2009; Luo et al. 2004). In the US juice industry alone, *Alicyclobacillus* spp. related spoilage is estimated to be around 35% (Chang and Kang, 2004). Recent reports suggest that there are intriguing linkages between the genotype of *Alicyclobacillus spp.* and what is found in specific fruit juice matrices (Zeki Durak et al., 2010). Thermal processes to inactivate the organisms are ineffective, and therefore spoilage control relies primarily on efficient detection of the organism (Groenewald et al., 2013). The International Federation of Fruit Juice Producers (IFU) recommended method of conventional plating requires a minimum of 72 hours of enrichment prior to plating. Thus, conventional plating methods to detect this organism are generally very time consuming and ineffective as a quality

* Correspondence: s-pillai@tamu.edu
[1]National Center for Electron Beam Research, Department of Nutrition and Food Science, Texas A&M University, College Station, Room 418B, Kleberg Center, MS 2472, 77843-2472 Texas, USA
Full list of author information is available at the end of the article

control strategy (Luo et al., 2004; Lin et al., 2005). Considering the high product value of juices, storing fruit juices for an extended period of time until obtaining conventional plating results is not economically feasible (Lin et al., 2005). A variety of rapid detection methods for *Alicyclobacillus spp.* and other organisms have have been reported in the literature (Al-Qadiri et al., 2006; Luo et al. 2004, Jasson et al., 2009, Cagnasso et al., 2010, Connor et al., 2005, Chang and Kang 2004). Many of the published reports have, however, focused only on limited types of samples (Luo et al 2004; Connor et al., 2005). Real-time PCR-based kits which have been tested for their specificity and sensitivity are currently commercially available. These assays rely on at least a 48 h enrichment step.

The underlying hypothesis of this study was that fruit juice and puree characteristics will influence the enrichment requirements of this organism, and that in some matrices the organism can be detected within 24 hours even when present at very low initial contamination. Therefore, the objective of this study was to compare the detection of the commonly encountered spoilage-related species, *Alicyclobacillus acidoterrestris*, in different fruit matrices after specific enrichment time frames using a commercially available real-time PCR kit compared to the IFU method. We tested 13 different retail store purchased fruit juice and puree matrices (varying in soluble solids content) that were spiked in the laboratory with low levels (10 CFU/ml) of *A. acidoterresteris* spores. Previous studies have employed higher inoculation concentrations such as 10^4 CFU/ml and 10^5 CFU/ml (Yamazaki et al., 2008; Wang et al., 2013, (Funes-Huacca et al., Funes-huacca et al. 2004). We chose this low level since a previous study suggested that *A. acidoterrestris* contamination concentrations in US juices could be low (Pettipher et al., 1997).

Methods
Sample juice matrices
Thirteen different fruit juice and purees (that varied in pH and Brix values) were employed in this study (Table 1). The samples were purchased from a local retail store and used within 48 h for all testing purposes, all prior to their expiry dates. The samples were opened under aseptic conditions (in a biosafety cabinet) to avoid any potential cross-contamination. Samples were taken for pH and Brix value determination. Digital hand-held refractometers (Reichert 35H and 65H, Depew, NY, USA) calibrated with distilled water were used for obtaining the Brix values.

Screening for background *Alicyclobacillus* contamination
The samples were initially screened for background *Alicyclobacillus* contamination per the International Federation

of Fruit Juice Producers method (IFU, 2007). The samples were heat-treated to 80°C for 10 min to eliminate background vegetative microbial populations and to enhance culturability of background *Alicyclobacillus* spp. spores. Portions (0.1 ml) of the samples were plated on K agar (HiMedia Laboratories, Mumbai, India) for enumerating background *Alicyclobacillus* spp. The plates were incubated for up to 5 days at 45°C per the IFU recommendations (IFU, 2007). The K agar plates were scored for growth/no-growth for *Alicyclobacillus* characteristic colonies. A commercial guaiacol detection kit (Kyokuto Pharmaceutical Co., Ltd., Japan) was used to confirm the presence of *Alicyclobacillus* in these colonies. The protocol provided by the kit manufacturer was followed to confirm the presence of guaiacol production by the colonies. The characteristic brown color change, as compared to the control blank samples, was used as an indicator for positive guaiacol formation.

Bacterial strain and growth conditions
An *Alicyclobacillus acidoterrestris* strain (ATCC 49025) was obtained from the American Type Culture Collection (Microbiologics®, USA), and cultured at 48°C in BAT broth (HiMedia Laboratories, Mumbai, India) for 5 days to obtain a cell titer of 10^7 CFU/ml. The culture was diluted in BAT broth to serve as the spiking inoculum.

Spiking of juice and puree samples
Aliquots (1 ml) of the juice and puree samples were aseptically pipetted (in triplicate) into separate 15 ml sterile conical tubes. Sterile BAT broth (9 ml) was added and the tubes were heated to 80°C for 10 min. To this, 0.1 ml of inoculum (containing 10 CFU of *A. acidoterrestris* cells) was added in triplicate. The inoculated samples were incubated in a shaking incubator at 45°C and 30 rpm for up to 48 h. Aliquots (1 ml and 0.1 ml) were taken at 24 hours and 48 hours from the inoculated samples. These samples were used for plating (0.1 ml) on K agar (Bevilacqua et al., 2008) and for DNA extraction (1 ml) for the real-time PCR assay. The "0 hour" sample was obtained from the heat-treated un-spiked sample.

DNA extraction and real-time PCR amplification
The commercial kits employed for DNA extraction and real-time PCR detection of *Alicyclobacillus* spp. and *A. acidoterrestris* in this study were the **food**proof® Short-Prep II Kit and the **food**proof® *Alicyclobacillus* Detection Kit (BIOTECON Diagnostics, Germany) respectively. The **food**proof® *Alicyclobacillus* Detection Kit uses 5' nuclease, Taqman (hydrolysis) probe chemistry to separately detect *Alicyclobacillus* genus (HEX channel) and *A. acidoterrestris* (FAM channel) as well as an internal amplification control (ROX channel) in one single assay. The kit manufacturer reports both *Alicyclobacillus* and

Table 1 - The effect of enrichment time on the Ct values of *Alicyclobacillus* and *A. acidoterrestris* in different juice and puree samples by real-time PCR

Sample	^{0}Brix	pH	*Alicyclobacillus* spp. (Ct values)			*A. acidoterrestris* (Ct values)		
			Background	24 h	48 h	Background	24 h	48 h
Orange concentrate	43	4.09	30.8 ± 3.1	33.9 ± 2.2[a]	34.3 ± 2.2[a]	ND	36.2 ± 2.4[a]	34.7 ± 2.7[a]
Apple concentrate	40	4.00	33.8[a] ± 3.9	47.7 ± 3.9[a]	40.2 ± 3.1[a]	32.3 ± 0.0	34.9 ± 1.5[a]	34.8 ± 1.7[a]
Ketchup	27	3.69	46.8 ± 5.5	46.3 ± 5.5[a]	39.3 ± 2.7[a]	ND	35.7 ± 4.4[a]	35.3 ± 0.5[a]
Tomato paste	26	3.69	39.0 ± 0.8	29.6 ± 11.6[a]	23.4 ± 13.7[a]	ND	25.6 ± 6.8[a]	18.9 ± 10.2[a]
Apple puree	17	4.08	ND*	34.0 ± 2.2[a]	21.4 ± 2.2[b]	ND	29.0 ± 4.8[a]	20.3 ± 9.9[a]
Peach puree	17	4.08	ND	UD	27.8 ± 2.7	ND	38.0 ± 1.6[a]	25.3 ± 2.1[a]
Apple juice (filtered)	12	3.8	27.1 ± 3.9	29.2 ± 2.2[a]	12.4 ± 2.7[b]	35.5 ± 0.9	26.6 ± 3.4[a]	10.4 ± 2.1[b]
Peach nectar	12	3.75	49.4 ± 5.5	19.8 ± 2.2[a]	14.3 ± 2.2[a]	35.4 ± 0.6	17.0 ± 2.10[a]	11.6 ± 0.2[a]
Apple juice (unfiltered)	11	3.73	37.0 ± 3.9	26.6 ± 2.2[a]	13.6 ± 2.2[b]	35.1 ± 0.7	23.9 ± 8.1[a]	11.0 ± 0.3[b]
Orange juice (with pulp)	11	4.03	38.2 ± 3.1	31.8 ± 2.4[a]	22.7 ± 2.2[b]	ND	26.7 ± 6.7[a]	20.6 ± 6.4[a]
Orange juice (filtered)	10	4.06	40.0 ± 5.5	22.5 ± 3.1[a]	19.3 ± 2.2[a]	35.2 ± 0.1	26.1 ± 8.3[a]	17.8 ± 7.7[b]
Tomato sauce	9	4.24	ND	31.7 ± 3.1[a]	16.3 ± 2.7[b]	37.0[a] ± 0.2	30.4 ± 3.8[a]	16.0 ± 3.0[b]
Tomato juice	6	4.15	37.6 ± 2.4	26.8 ± 1.7[a]	15.7 ± 1.7[b]	33.4 ± 1.7	26.1 ± 3.9[a]	14.1 ± 6.6[b]

Ct values with different superscript letters within each row indicate statistically ($p < 0.01$) significant differences as a function of enrichment time. Pairwise comparisons were performed using the Student's *t* test.
*ND = Not detected.

A. acidoterrestris amplification chemistries have a limit of detection of 7 genome equivalents, and 10^1-10^2 CFU/ml in combination with the DNA extraction kit. However, the manufacturer's verification documentation demonstrates detection at even 1 genome equivalent, though less than 100% (*data not included*). Though specificity testing was not a goal of this paper, the kit manufacturer's verification documentation states that 100% of 40 strains of non-target organisms including related genera (e.g., *Bacillus spp.*, *Listeria spp.*) as well as strains commonly found in food were properly excluded (tested negative) (*data not included*).

When performing DNA extraction and real-time PCR, the kit manufacturer's recommended protocols were followed. Briefly, 1 ml from the 0 h (un-spiked), 24 h enriched (inoculated) and 48 h enriched (inoculated) samples were aseptically taken and used for DNA extraction. The un-spiked sample was included to identify possible natural contamination of the samples. For the real-time PCR assay, 5 μl aliquots of the extracted DNA samples were used as template according to the manufacturer's protocol. The real-time PCR amplifications were performed using duplicate technical replicates. The real-time PCR reaction mixture contained 20 μl of the premixed commercial PCR reagent (18.0 μl master mix, 1.0 μl enzyme solution and 1.0 μl internal amplification control) and 5 μl genomic DNA extract. For the negative and positive controls, 5 μl PCR-grade H_2O and 5 μl **food**proof® *Alicyclobacillus* Control Template respectively (provided in the kit) were added to the premixed commercial PCR reagent. Following the **food**proof® *Alicyclobacillus*

Detection Kit protocol, the real-time PCR amplification parameters used in the Applied Biosystems 7900HT real-time PCR cycler were as follows: 37°C for 4 min (1 cycle) and 95°C for 5 min (1 cycle), and 95°C for 5 s followed by 60°C for 1 min (50 cycles). The enzyme solution of the **food**proof® *Alicyclobacillus* Detection Kit contains not only Taq polymerase, but also uracyl-N-glycosylase (UNG). The UNG is included to prevent false-positive reactions from previous PCR reactions as the dNTPs in the master mix contain dUTP rather than dTTP. Thus, should cross-contamination occur from previously made amplificates containing dUTP, they will be degraded by UNG during the initial 37°C incubation at the start of the PCR program. The inbuilt cycler software (SDS V.2.4, 2009, California, USA) was used for obtaining the real-time PCR cycle threshold (Ct) values.

Statistical analysis

The experiments were performed using three biological replicates for the fruit matrix samples with each of the real-time PCR amplifications performed in duplicate. The real-time PCR data as presented is the mean of six Ct values ± standard deviation. The correlation between Brix values and Ct values were determined using the Pearson product Moment Correlation using SigmaPlot (Systat Software, Inc.). To discern the effect of juice matrix, incubation time and the possible interaction between the juice matrix and time of incubation the Ct values were evaluated using ANOVA at a 1% level of significance using the commercially available Design-Expert™ (Ver 7.1.4) software. Significant differences between the treatments if any, were determined using Student's *t*-test at a 1% level of significance (P < 0.01).

Results and Discussion

The 13 fruit juice and puree samples in the study had quite similar pH values which ranged from 3.69 to 4.24. However, they varied considerably in terms of their Brix values. Tomato juice had the lowest Brix value of 6.0, while the orange concentrate had the highest value with 43 (Table 1). The real-time PCR kit used provided separate amplification plots for the genus *Alicyclobacillus* and the species *A. acidoterrestris* as they were detected using different dyes. The Ct value in the real-time PCR assays was used as the benchmark for "detection". Ct values greater than 50 were scored (by the instrument software) as "not-detected". All samples, except for apple puree, peach puree, and tomato sauce showed the presence of background *A. acidoterrestris* contamination. Ten out of 13 samples (77%) showed background contamination of the genus *Alicyclobacillus*, while 7 out of the 13 samples (54%) showed the specific presence of *A. acidoterrestris* background contamination (Table 1). A number of reports have shown that *A. acidoterrestris* can be present in both fresh juices and juice concentrates (Pettipher et al. 1997; Luo et al., 2004; Groenewald et al. 2009; Groenewald et al. 2013; McKnight et al., 2010; Wang et al., 2013). The presence of background *Alicyclobacillus* spp. but not *A. acidoterrestris*, such as in orange concentrate, suggests that there may have been species other than *A. acidoterrestris* in this sample.

Based on the Ct values, it is evident that the background contamination levels of the target organisms varied in the samples (Table 1). Filtered apple juice had the highest background levels of *Alicyclobacillus* spp. contamination (Ct value =27.1) while ketchup had the lowest levels (Ct value = 46.8 ± 5.5). In terms of *A. acidoterrestris* background levels, apple concentrate had the highest levels (Ct value = 32.3) compared to tomato sauce which had the lowest levels (Ct value = 37.0 ± 0.2). There was only one sample (tomato sauce) where the assay detected the background presence of *A. acidoterrestris* but not the genus *Alicyclobacillus*. This result may be related to juice matrix components preventing effective amplification of the genus-specific sequences (compared to species-specific sequences) when these target sequences are present in low levels. This is further supported by lower Ct values for background levels of *A. acidoterrestris* compared to background levels of the genus *Alicyclobacillus* in the apple concentrate, peach nectar, apple juice, orange juice, and tomato juice samples (Table 1).

The **food**proof® *Alicyclobacillus* Detection Kit was designed to have a higher preference for the FAM-targeted *A. acidoterrestris* species amplification chemistry over the HEX-targeted *Alicyclobacillus* genus amplification chemistry. This preference is important at low DNA concentrations because if the species is found present, then naturally the genus is considered present as well. However, without this preference, the more specific *A. acidoterrestris* species information could be lost.

Table 1 also illustrates the effect enrichment time has on the ability of a commercial real-time PCR assay to detect *Alicyclobacillus* genus-specific sequences and *A. acidoterrestris* species-specific sequences when spiked at low initial concentrations (10 CFU/ml). When the samples were inoculated with a low inoculum of 10 CFU/ml, all samples (except for peach puree) showed detectable *Alicyclobacillus* spp. within 24 hours. By 48 hours of enrichment, all samples were "positive" for *Alicyclobacillus* spp. Within 24 h of enrichment, all 13 samples showed the presence of A. *acidoterrestris* (Table 1). The ability to detect *A. acidoterrestris* in all 13 samples after 24 hours of enrichment, even in peach puree, supports the finding of higher efficiency for the species-specific primer sequences compared to the genus-specific sequences. The variability in Ct values between 24 h and 48 h enrichments appears to be due to differences in the growth rate of *Alicyclobacillus spp.* within the different sample matrices. Previous studies have shown juice matrices to differentially influence *A. acidoterrestris* growth likely due to pH, total soluble solids, water activity, etc. (Silva et al., 1999; Bahceci and Acar, Bahçeci and Acar 2007; Collins, 2008; Goto et al., 2007). In a recent study by Wang et al. (2014), real-time PCR detection was coupled with immuno-magnetic separation (IMS) to detect low levels (10 CFU/ml) of *Alicyclobacillus* spp as well. They, however, state that in complex matrices there was low correlation between *A. acidoterrestris* numbers and Ct values. They attributed this low correlation to interfering sample debris. The influence of sample matrix effects can be deduced from the *A. acidoterrestris* Ct values of "filtered" samples and "unfiltered samples" (Table 1). In filtered orange juice there appears to be inhibition of *A. acidoterrestris* growth; hence, there is no significant difference (p > 0.05) in the Ct values between the 24 h and 48 h enriched samples. However, in orange juice "with pulp" samples, there was a significant difference (P < 0.05) between the Ct values reported after 24 h and 48 h enrichment, suggesting pulp may hinder the growth rate of *A. acidoterrestris* in such samples.

The analysis of variance indicates that both incubation time and the fruit sample matrix have a statistically significant (p < 0.01) effect on the Ct values for *A. acidoterrestris* detection (Table 2). Sample texture also appears to have a significant effect on Ct values. This is evident in the statistically significant differences in the Ct values for *A. acidoterrestris* detection after 24 h and 48 h enrichment in the tomato-based samples, i.e., tomato sauce, tomato paste and ketchup samples. Similar differences were observed in the apple-based samples, i.e., apple juice and apple puree

Table 2 Analysis of variance (ANOVA) to determine the interaction of sample type, enrichment time, and real-time PCR detection (Ct value) of *A. acidoterrestris*

Source	Sum of squares	df	Mean square	F Value	p-value
Model	9302.70	23	404.47	10.09	<0.0001
Sample type	6280.43	12	523.37	13.06	<0.0001
Enrichment time	2866.86	1	2866.86	71.53	< 0.0001
Sample type *X* enrichment time	1148.67	10	114.87	2.87	0.0035

samples (Table 1). The Pearson Product Moment Correlation analysis showed the Ct value for *A. acidoterrestris* detection tended to increase with increasing Brix values (correlation coefficient: 0.608; P value 0.02); suggesting that the higher soluble solids content suppressed the growth (i.e., higher Ct values) of the organism in the enrichment medium. These results are supported by previous reports which suggest that *A. acidoterrestris* growth is negatively influenced at Brix values above 18-20 (Collins, 2008; Sprittstoesser et al., 1994). In this study using commercially available kits, we found that these effects began at Brix values of greater than 27.

Table 3 illustrates the enrichment time requirements for the detection of *A. acidoterrestris* (by real-time PCR and growth on K agar) and the confirmation of guaiacol formation. For all 13 samples that were tested, real-time PCR was capable of detecting low initial contamination (10 CFU/ml) of *A. acidoterrestris* within 24 h of enrichment by real-time PCR. However, two samples (orange concentrate and ketchup) were negative even after 48 h of enrichment based on growth on K agar. All samples that showed positive growth on K agar after 24 h or 48 h enrichment were also positive for guaiacol

formation (Table 3). There did not appear to be any pattern in culture-based detection or guaiacol formation as a function of the Brix values. Instead, having a minimum enrichment time of 48 hours was very important for the conventional IFU plating method to be effective. This study made use of the most commonly identified spoilage species *A. acidoterrestris*, which we exposed to a period of heat-stress for 10 minutes. Similar studies should be performed with other pertinent *Alicyclobacillus* species in samples that have been exposed to different processing conditions. Nevertheless, this study provides evidence that faster more accurate detection of the genus *Alicyclobacillus* as well as the species *A. acidoterrestris* is possible and should be further explored by fruit processing manufacturers to best optimize detection in their particular matrices.

Conclusions

Fruit juice and puree characteristics, specifically the soluble solids content (Brix values), significantly influence the enrichment and real-time PCR detection of the spoilage genus *Alicyclobacillus* as well as the species *A. acidoterrestris*. Samples with high Brix values impair

Table 3 Comparison of detection of *A. acidoterrestris* in different fruit matrices by real-time PCR and conventional plating using K agar and confirmation of guaiacol formation

Sample	°Brix	Time requirement*	Growth on K Agar		Guaiacol Formation	
			24 h	48 h	24 h	48 h
Orange concentrate	43	24 h	-	-	-	-
Apple concentrate	40	24 h	+	+	+	+
Ketchup	27	24 h	-	-	-	-
Tomato paste	26	24 h	+	+	+	+
Apple puree	17	24 h	-	+	-	+
Peach puree	17	24 h	-	+	-	+
Apple juice (filtered)	12	24 h	-	+	-	+
Peach nectar	12	24 h	+	+	+	+
Apple juice (unfiltered)	11	24 h	+	+	+	+
Orange juice (with pulp)	11	24 h	-	+	-	+
Orange juice (no pulp)	10	24 h	-	+	-	+
Tomato sauce	9	24 h	+	+	+	+
Tomato juice	6	24 h	+	+	+	+

*enrichment time required for positive detection by real-time PCR.
-: not detectable.

microbial proliferation in the enrichment medium. The impaired microbial growth results in higher Ct values in real-time PCR assays. Thus, fruit juices and purees with high Brix values require longer enrichment periods or need to be diluted prior to being enriched for real-time PCR detection. This study also highlights the point that even with low starting concentrations namely 10 CFU/ml, the commercial real-time PCR kit used detected *A. acidoterrestris* within 24 h of enrichment. Thus, it may not be necessary to wait for 48 h of enrichment prior to real-time PCR analysis. The ability to detect the presence of this key spoilage organism in fruit juices and purees within 24 h will have significant economic value to the fruit juice industry.

Competing interests

There are no personal or financial competing interests in this study. SS and SDP are employees of the State of Texas. CH is an employee of BIOTECON Diagnostics which provided the molecular reagents for this study.

Authors' contributions

SS, SDP, and CH were involved in the experimental design of this study. SS and SDP performed the laboratory analyses. SS, SDP and CH were involved in the writing, editing, and proof-reading of the manuscript.

Acknowledgements

The authors wish to acknowledge Drs. C. Groenewald and K. Berghof-Jaeger of BIOTECON Diagnostics for their support and for kindly providing the real-time PCR reagents used in this study. This work was supported by funding to SDP from the Texas A&M AgriLife Research project H8708.

Author details

[1]National Center for Electron Beam Research, Department of Nutrition and Food Science, Texas A&M University, College Station, Room 418B, Kleberg Center, MS 2472, 77843-2472 Texas, USA. [2]BIOTECON Diagnostics GmbH, Hermannswerder 17, 14473 Potsdam, Germany.

References

Al-Qadiri L, Cavinato R (2006) Fourier transform infrared spectroscopy, detection and identification of *Escherichia coli* O157:H7 and *Alicyclobacillus* strains in apple juice. Int J Food Microbiol 111(1):73–80

Bahçeci KS, Acar J (2007) Modeling the combined effects of pH, temperature and ascorbic acid concentration on the heat resistance of *Alicyclobacillus acidoterrestris*. Int J Food Microbiol 120:266–73

Bahçeci KS, Serpen A, Gökmen V, Acar J (2005) Study of lipoxygenase and peroxidase as indicator enzymes in green beans: change of enzyme activity, ascorbic acid and chlorophylls during frozen storage. J Food Eng 66:187–92

Bevilacqua A, Sinigaglia M, Corbo MR (2008) *Alicyclobacillus acidoterrestris*: New methods for inhibiting spore germination. Int J Food Microbiol 125:103–10

Cagnasso S, Falasconi M, Previdi MP, Franceschini B, Cavalieri C, Sberveglieri V, Rovere P (2010) Rapid screening of *Alicyclobacillus acidoterrestris* spoilage of fruit juices by electronic nose: A confirmation study. J Sensors. doi:10.1155/2010/143173

Chang SS, Kang DH (2004) Alicyclobacillus spp. In the fruit juice industry: history, characteristics, and current isolation/detection procedures. Critical Rev Microbiol 30:55–74

Collins J (2008) *Alicyclobacillus* best practice guideline, A guideline for the reduction and control of thermophylic, sporeforming bacteria (Alicyclobacillus species, ACB) in the production, packing and distribution of fruit juices, juice concentrates purees and nectars, ACB Best Practice Guideline, July 2008 – AIJN, 3-29

Connor CJ, Luo H, Gardener BBM, Wang HH (2005) Development of a real time PCR-based system targeting the 16S rRNA gene sequence for rapid detection of *Alicyclobacillus spp*. In juice products. Int J Food Microbiol 99 (3):229–35

Eguchi S, Pinhatti MEC, Azuma EH, Manfio GP, Canhos VP (2000) An ecological study of acidothermophilic sporulating bacteria (Alicyclobacillus) in the citrus industry. In: Annals of the 23rd IFU Symposium. Havana, Cuba, pp 257–70

Funes-huacca M, Correia De Almeida Regitano L, Mueller O, Carrilho E (2004) Semiquantitative determination of *Alicyclobacillus acidoterrestris* in orange juice by reversetranscriptase polymerase chain reaction and capillary electrophoresis – laser induced fluorescence using microchip technology. Electrophoresis 25(21-22):3860–4

Gocmen D, Elston A, Williams T, Parish M, Rouseff RL (2005) Identification of medicinal off-flavours generated by *Alicyclobacillus* species in orange juice using GC-olfactrometry and GC-MS. Lett App Microbiol 40:172–7

Goto K (2007) Parameters for detection of *Alicyclobacillus* and test methods. In: Yokota A, Fujii T, Goto K (eds) *Alicyclobacillus* Thermophilic Acidophilic Bacilli. Springer, New York

Groenewald WH, Gouws PA, Witthuhn RC (2009) Isolation, identification and typification of *Alicyclobacillus acidoterrestris* and *Alicyclobacillus acidocaldarius* strains from orchard soil and the fruit processing environment in South Africa. Food Microbiol 26:71–6

Groenewald WH, Gouws PA, Witthuhn RC (2013) Thermal inactivation of *Alicyclobacillus acidoterrestris* spores isolated from a fruit processing plant and grape concentrate in South Africa. Afr J Microbiol Res 7(22):2736–40

IFU (2007) IFUMB12. Method on the detection of taint producing *Alicyclobacillus* in fruit juices. http://www.ifu-fruitjuice.com/ifu-methods.

Jasson V, Rajkovic A, Baert L, Debevere J, Uyttendaele M (2009) Comparison of enrichment conditions for rapid detection of low numbers of sublethally injured Escherichia coli O157 in food. J Food Prot 72:1862–8

Lin M, Al-Holy M, Chang SS, Huang Y, Cavinato AG, Kang DH, Rasco BA (2005) Rapid discrimination of *Alicyclobacillus* strains in apple juice by Fourier transform infrared spectroscopy. Int J Food Microbiol 105:369–76

Luo H, Yousef AE, Wang HH (2004) A real-time polymerase chain reaction-based method for rapid and specific detection of spoilage *Alicyclobacillus* spp. in apple juice. Lett App Microbiol 39(4):376–82. doi:10.1111/j.1472-765X.2004.01596.x

McKnight IC, Eiroa MN, Sant'Ana AS, Massaguer PR (2010) *Alicyclobacillus acidoterrestris* in pasteurized exotic Brazilian fruit juices: isolation, genotypic characterization and heat resistance. Food Microbiol 27(8):1016–22. doi:10.1016/j.fm.2010.06.010

Pettipher GL, Osmundson ME, Murphy JM (1997) Methods for the detection and enumeration of Alicyclobacillus acidoterrestris and investigation of growth and production of taint in fruit juice and fruit juice-containing drinks. Lett Appl Microbiol 24:185–9

Silva FVM, Gibbs P (2001) *Alicyclobacillus acidoterrestris* spores in fruit products and design of pasteurization processes. Trends food sci Tech 12:86–74

Silva FM, Gibbs P, Vieira MC, Silva CLM (1999) Thermal inactivation of *Alicyclobacillus acidoterrestris* spores under different temperature, soluble solids and pH conditions for the design of fruit processes. Int J Food Microbiol 51:95–103

Sprittstoesser DF, Churey JJ, Lee CY (1994) Growth characteristics of aciduric sporeforming bacilli isolated from fruit juices. J Food Prot 57:1080–3

Walls I, Chuyate R (2000) Spoilage of fruit juices by *Alicyclobacillus acidoterrestris*. Food Australia 52(7):286–8

Wang Z, Wang J, Yue T, Yuan Y, Cai R, Niu C (2013) Immunomagnetic separation combined with polymerase chain reaction for the detection of *Alicyclobacillus acidoterrestris* in apple juice. PLoS One 8(12):82376. doi:10.1371/journal.pone.0082376

Yamazaki K, Teduka H, Inoue N, Shnano H (2008) Specific primers for detection of *Alicyclobacillus acidoterrestris* by RT-PCR. Lett Appl Microbiol 23(5):350–4

Zeki Durak M, Churey JJ, Danyluk MD, Worobo RW (2010) Identification and haplotypedistribution of *Alicyclobacillus* spp. from different juices and beverages. Int J Food Microbiol 142:286–91

Presence of arsenic in Sri Lankan rice

Channa Jayasumana[1*], Priyani Paranagama[2], Saranga Fonseka[2], Mala Amarasinghe[2], Sarath Gunatilake[3] and Sisira Siribaddana[4]

Abstract

Background: Arsenic and heavy metals are implicated in causation of CKDu among farmers in dry zone of Sri Lanka. Rice has been identified as a major source of arsenic in research carried out in other countries. We analyzed 120 samples of new improved varieties (NIVs) and 50 samples of traditional varieties (TV) of rice for total arsenic content.

Findings: Rice cultivated in Sri Lanka is contaminated with arsenic. Agrochemical dependent NIVs contain considerable amount (20.6 -540.4 µg/Kg) of arsenic. There is no difference between the arsenic content in NIV rice samples from areas where there is high or low prevalence of CKDu. TVs that are cultivated without using agrochemicals contain significantly less arsenic (11.6 - 64.2 µg/Kg). However, it is evident that the TVs also contain toxic metals if they are grown with fertilizers and pesticides.

Conclusion: A high proportion of arsenic in rice exists in the inorganic form. Sri Lanka is a nation with high per capita consumption of rice. Codex Alimentarius recommends the maximum allowable limit for inorganic arsenic in rice as 200 µg/kg. Assuming that 70% of the total arsenic content exists in the inorganic form, this corresponds to a level of about 286 µg/kg of total arsenic. As such, 11.6% of the samples of NIVs exceeded this maximum recommended level in polished rice. Inorganic arsenic is a non-threshold carcinogen. Research should be focused on developing rice varieties that do not retain arsenic within the rice grain.

Keywords: Arsenic; Rice; Chronic kidney disease; Sri Lanka; Agrochemicals

Findings

Arsenic (As) has been a known poison for thousands of years and is classified as a class one non-threshold human carcinogen (Europian Food Safety Authority 2009). Arsenic was not identified as an environmental pollutant in Sri Lanka until recently. Arsenic was identified as a possible etiological factor for the newly emerging epidemic of Chronic Kidney Disease of unknown origin (CKDu), a tubulo-interstitial nephritis among paddy farmers in dry zone of Sri Lanka (Jayasumana et al. 2013; 2014). Analytical studies have shown that a significant amount of arsenic in biological samples (urine, hair and nails) of these CKDu patients (Jayasumana et al. 2013; Jayatilake et al. 2013). Hence, we decided to analyze rice grown in Sri Lanka for arsenic. Analysis was carried out at the department of chemistry, University of Kelaniya and again at the Institute for Integrated Research in Materials, Environments and Society (IIRMES) lab, California State University, Long Beach (CSULB), USA.

A total of 170 rice samples were collected from seven different locations to polythene zip bags (Figure 1). There were reported cases of CKDu in and around the sample collection centers from Padaviya, Sripura and Mahawilachchiya but not in at other locations (Kurunegala, Mihinthale, Moneragala and Gampaha). Hundred and twenty of these samples were Newly Improved Varieties (NIV) of rice and were collected directly from paddy farmers. Other 50 samples consisted of 5 different types of traditional varieties (TV) of rice, all cultivated at Kurunegala and Sripura without using agrochemicals but with plant extracts for pest control and with the use of natural fertilizers. Six samples of TVs were purchased from a supermarket in Colombo. They were marketed by a private company and produced, using agrochemicals. Samples were ground using an agate mortar passed through 0.3 mm sieve before analysis. Atomic absorption spectrometry (AAS) with hydride generation system (GBC 3000) and background corrector (GBC 932 plus GBC

* Correspondence: jayasumanalk@yahoo.com
[1]Faculty of Medicine& Allied Sciences, Rajarata University of Sri Lanka, Saliyapura 50008, Sri Lanka
Full list of author information is available at the end of the article

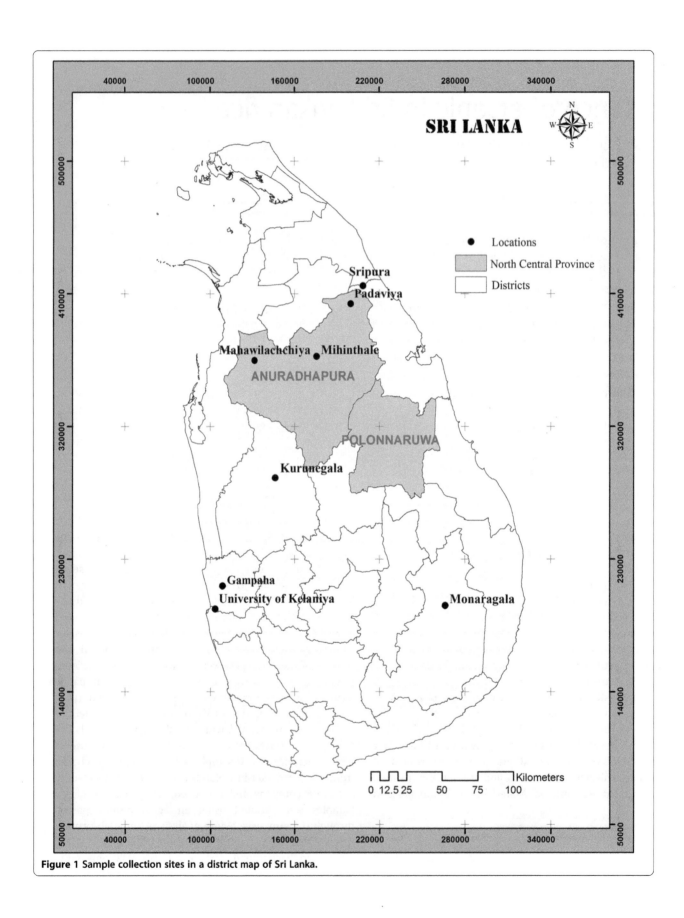

Figure 1 Sample collection sites in a district map of Sri Lanka.

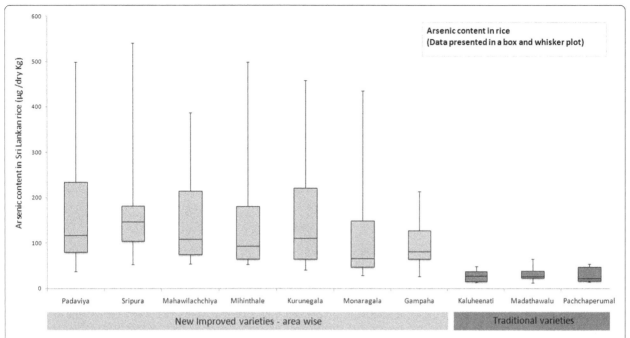

Figure 2 Arsenic content in Sri Lankan rice. Presented as Box and whisker plot (μg/dry Kg). *The box represents area between 1st and 3rd quartile with the center line representing the median. Whiskers represent maximum and the minimum value.

scientific equipment's, VIC, Australia) was used in University of Kelaniya to detect arsenic. Method detection limit (MDL) was 10 μg/Kg. Inductively Coupled Plasma Mass Spectrometer (ICP-MS; HP 4500, Agilent Technologies, Palo Alto, CA) equipped with a quadrupole analyzer and octopole collision/reaction cell was used to detect arsenic and other trace metals at IIRMES lab, CSULB (EPA method 6020 m). Mercury was detected by cold vapor atomic fluorescence spectroscopy using EPA method 245.7m. MDL was 1 μg/g for aluminum and iron, 0.01 μg/g for mercury and 0.025 μg/g for all other trace metals. Samples tested at IIRMES laboratory are not split samples of those tested in Sri Lanka. Data analysis was done using Microsoft excel 2007.

In 1998, rice has been identified as potentially important source of arsenic to humans for the first time (Yost et al.

1998) and is the largest dietary source of inorganic arsenic (Tsuji et al. 2007). Arsenic accumulation in rice is a newly recognized disaster for South-East Asia (Meharg 2004). Rice is particularly susceptible to arsenic accumulation compared to other cereals because it is grown anaerobically in paddy fields that are flooded (Meharg and Zhao 2012). Inorganic arsenic in the soil is inter-converted between the arsenite [As III] and arsenate [As V]. Arsenite predominates in anaerobic environments in contrast to arsenate, which is seen predominantly under aerobic conditions. However, organic monomethylarsonic acid (MMA) and dimethylarsinic acid (DMA) is also present in rice (Meharg and Hartley-Whitaker 2002). Around 50% of total arsenic in rice is inorganic and the remainder is DMA (Heikens 2006). In South Asian countries the percentage of toxic inorganic arsenic is relatively higher (Rahman and Hasegawa 2011).

Table 1 Arsenic content in Sri Lankan rice (μg/dry Kg)

	New improved varieties - area wise							Traditional varieties		
	Padaviya	Sripura	Maha Wilachchiya	Mihinthale	Kurunegala	Monaragala	Gampaha	Kalu Heenati	Mada Thawalu	Pachcha Perumal
	N = 20	N =17	N =25	N = 17	N = 19	N = 11	N = 11	N = 10	N = 10	N = 10
Minimum	37.2	52.4	54.2	52.4	40.2	28.8	26.6	12.8	11.6	14.4
Maximum	498.6	540.4	387.6	498.4	458.4	435.2	212.8	48.6	64.2	54.2
Mean	162.8	186.1	144.3	141.2	152.6	129.3	96.2	28.1	30.5	28.5
Median	116.8	146.2	108.4	92.5	110.6	65.4	80.6	27.6	26.7	22.4
Daily As intake from rice (μg)	51.7	59.1	45.6	44.9	48.5	41.1	30.6	8.9	9.7	9.1

Table 2 Trace metal values of traditional varieties of rice-ICP-MS analysis (µg/dry g) please note that the units are different from Table 1

	Al[1]	Sb	As[1]	Ba	Be	Cd[1]	Cr[2]	Co	Cu[2]	Fe[2]	Pb[1]	Mn[3]	Hg[1]	Mo[2]	Ni[3]	Se[2]	Ag	Sr	Tl	Sn[1]	Ti	V[2]	Zn[2]
Kalu Heenati (n = 6)	4.3	0.1	ND	0.7	ND	ND	0.7	0.3	2.3	7.7	ND	9.3	ND	0.7	1.3	0.2	0.2	0.3	ND	0.1	0.5	1.0	11.6
Kuruluthuda (n = 4)	1.5	0.1	ND	0.6	ND	ND	0.7	0.3	1.8	3.1	ND	5.2	ND	0.4	0.6	0.2	ND	0.2	ND	0.1	0.3	0.9	7.15
Madathawalu (n = 2)	3.3	0.1	ND	0.6	ND	ND	0.7	0.3	2.4	4.8	ND	8.0	ND	0.4	1.4	0.2	ND	0.2	ND	0.1	0.3	0.9	10.9
Kahawanu (n = 2)	ND	0.1	ND	0.3	ND	ND	0.7	0.3	2.2	2.1	ND	4.0	ND	0.6	0.8	0.2	ND	0.1	ND	0.1	0.2	1.0	6.45
Kalu Heenati* (n = 2)	2.0	0.1	0.1	1.0	ND	ND	0.8	0.3	2.0	12.0	ND	16.9	ND	0.6	2.5	0.2	ND	0.3	ND	0.1	0.6	1.2	17.0
Madathawalu* (n = 2)	2.4	0.1	0.1	3.4	ND	ND	0.6	0.3	1.8	7.8	ND	17.0	ND	0.7	1.4	0.3	ND	0.4	ND	0.1	0.4	1.0	17.3
Pachchaperumal* (n = 2)	7.5	0.1	0.1	1.2	ND	ND	0.7	0.4	2.3	7.4	ND	11.1	ND	0.4	9.5	0.3	ND	0.8	ND	0.1	0.5	1.0	12.9

[1]Toxic elements [2]Essential elements and trace elements (in overdose some may be toxic)) [3]Probably essential elements. (http://apps.who.int/iris/bitstream/10665/37931/1/9241561734_eng.pdf).
*non-organically cultivated.

Rice is the staple food of majority of Sri Lankans and is grown widely in the dry zone. In 2010 annual per capita rice consumption in Sri Lanka was 116 Kg (http://www.agridept.gov.lk/images/stories/site/PDF/Publication/English/BOOK/proposedplan.pdf). In the 1960s the International Rice Research Institute based in Philippines, introduced the new improved varieties (NIV). Development of Bg34-8 at Bathalagoda- a major rice research institute in Sri Lanka with a yield potential of 7 t/ha became very popular, replacing the widely grown traditional variety Pachchaperumal. Within 60 years, NIVs have replaced TVs making them almost extinct. Nevertheless, research has shown that TVs are richer in protein, iron, antioxidants, anti-amylase, anti-glycation and glycation reversing activities in comparison to the NIVs (Gunaratne et al. 2013; Premakumara et al. 2013).

We were unable to perform arsenic speciation due to non-availability of facilities and the lack of funds. However, arsenic speciation is important as it varies with type of rice (Booth 2008). According to the Codex Alimentarius, the maximum allowable limit for inorganic arsenic in rice is 200 µg/kg (http://www.fao.org/news/story/en/item/238802/icode/). Assuming that 70% of the total arsenic content exists in the inorganic form, this corresponds to a level of about 286 µg/kg of total arsenic. As such, 11.6% of the samples of NIVs exceeded this maximum recommended level in polished rice. Person (60 kg of body weight) eating rice obtained from Sripura area consumes approximately 1 µg/kg/day of arsenic. Assuming 70% of this is inorganic arsenic the daily exposure would approximate to 0.7 µg/kg. Although this value exceeds the United States environmental protection agency reference (0.3 µg/kg/day) (http://www.epa.gov/teach/chem_summ/Arsenic_summary.pdf) it is below the value recommended by joint FAO/WHO expert committee on food additives (2-7 µg/kg/day) for the avoidance of potential cancers of skin, lung and bladder (http://www.who.int/ipcs/features/arsenic.pdf). Highest amount of arsenic (540.4 µg/kg) was reported in a sample obtained at Sripura, a farming colony where the CKDu epidemic originated in the 1990s. (Table 1 and Figure 2).

Our measurements reflected the total arsenic content (both inorganic and organic) and the tested NIVs were polished rice but TV were rice with bran. The Codex Alimentarius standard is established for polished rice and not for brown rice. Rice bran was found to contain approximately seven to nine time higher concentrations of total and inorganic arsenic respectively than those found in the corresponding polished rice (Narukawa et al. 2014; Ruangwises et al. 2012). Our results are similar to those of Chandrajith et al. (2011) who reported an arsenic content of (90–260 µg/kg) in an analysis of 10 samples of rice done by ICP-MS in two other endemic areas for CKDu. However, an earlier analysis done by Instrumental Neuron Activation followed by high-resolution Υ-ray spectrometry which revealed an arsenic content of 34–92 µg/kg in rice (Jayasekera and Freitas 2005). There is no difference between the arsenic content in NIVs from the dry zone and wet zone (Gampaha). Compared to NIVs, arsenic content in TVs cultivated without agrochemicals is low, when tested with AAS, (Table 1). These levels are lower than the MDL when tested with ICP-MS method (Table 2). MDL for arsenic in ICP-MS was 25 µg/Kg. However, the TVs cultivated using agrochemicals contains 100 µg/Kg of arsenic (Table 2). Furthermore, a higher amount of Fe, Mn, Ni and Zn were also detected in non-organically cultivated TVs than the organically cultivated TVs. Out of the TVs, Madathawalu showed the highest amount of Ba, Mn and Zn while Kalu Heenati contained the highest amount of iron. None of the tested TVs contained Be, Cd, Pb, Hg and Tl.

These results imply that the arsenic in rice grown in Sri Lanka, originates most probably from the agrochemicals. We have already identified phosphate fertilizers as a main source of arsenic in Sri Lanka (Fernando et al. 2012). Arsenic accumulates in soil following repeated long-term application of contaminated fertilizers. Arsenic in TVs of rice produced without agrochemicals may originate from already contaminated soil. Farmers started the cultivation of TVs again without agrochemicals only two seasons prior to our study. Arsenic content in Sri Lankan rice is not as high as in West Bengal, Bangladesh and other East Asian countries (Zhu et al. 2008; Rahman and Hasegawa 2011). However, there is no CKDu similar to Sri Lanka in these countries. Therefore, arsenic in rice cannot be linked directly to the CKDu although the additive effects of arsenic in combination with other heavy metals cannot be ruled out. However, given the high per capita consumption, the adverse effects due to arsenic in rice cannot be overlooked, as it is a non-threshold carcinogen and is linked to many non-communicable diseases (Tchounwou et al. 2003; Kapaj et al. 2006). Jasmine rice from Thailand and Basmati rice (both TVs) from South Asia contain the least amount of arsenic (Potera 2007). The arsenic in the rhizosphere is absorbed through ubiquitous aquaglycoproteins and phosphate channels in the rice roots (Mukhopadhyay et al. 2014). Activity of these proteins and channels and many other factors determines whether rice plant is hyper-tolerant or a hyper-accumulator and whether the rice grain contains excessive amounts of arsenic or not (Srivastava et al. 2012).

Water management and rhizosphere manipulation can alter the arsenic concentration in the rice (Meharg and Zhao 2012). Arsenic content in the rice also can be reduced by rinse washing and cooking it in large amount of water (Raab et al. 2009).

Abbreviations
CKDu: Chronic kidney disease of unknown origin; NIV: New improved varieties; TV: Traditional varieties; MDL: Method detection limit; MMA: Monomethylarsonicacid; DMA: Dimethylarsinic acid.

Competing interests
The authors declare that they have no competing interests.

Author's contributions
CJ, PP, MA and SF designed and performed the experiment; CJ, SS and SG analyzed the data; CJ and SS wrote the first draft of the manuscript. All authors read and approved the final manuscript.

Acknowledgements
Contribution of Dr. Chinthaka Wijewardhane and Manjula Ranagalage is greatly acknowledged. Project on Higher Education for Twenty first Century (HETC) of the University Grants Commission and Farmers trust fund of Sri Lanka provided the financial support. IIRMES lab director Richard Gossett and his staff extended their maximum support for the analytical study. We are extremely grateful for the expert comments by two anonymous reviewers that helped to enhance the quality of the manuscript.

Author details
[1]Faculty of Medicine& Allied Sciences, Rajarata University of Sri Lanka, Saliyapura 50008, Sri Lanka. [2]Faculty of Science, University of Kelaniya, Colombo 11600, Sri Lanka. [3]Department of Health Science, California State University Long Beach, Long Beach, CA 90840, USA. [4]Faculty of Medicine & Allied Sciences, Rajarata University of Sri Lanka, Saliyapura 50008, Sri Lanka.

References
Booth B (2008) Arsenic speciation varies with type of rice. Environ Sci Technol 42:3484–3485

Chandrajith R, Nanayakkara S, Itai K et al (2011) Chronic kidney diseases of uncertain etiology (CKDue) in Sri Lanka: geographic distribution and environmental implications. Environ Geochem Health 33:267–78

Europian Food Safety Authority (2009) Scientific Opinion on Arsenic in Food 1. EFSA J 7:1351

Fernando A, Jayalath K, Fonseka SI, Jayasumana MACS, Amarasinghe MD, Senanayake VK, Kannangara A, et al. (2012) Determination of Arsenic content in synthetic and organic manure based fertilizers available in Sri Lanka. In Proceedings of the International Conference on Chemical Sciences, Institute of Chemistry, Ceylon Colombo, Sri Lanka, 129–134

Gunaratne A, Wu K, Li D et al (2013) Antioxidant activity and nutritional quality of traditional red-grained rice varieties containing proanthocyanidins. Food Chem 138:1153–1161

Heikens A (2006) Arsenic contamination of irrigation water, soil and crops in Bangladesh: risk implications for sustainable agriculture and food safety in Asia. 1st edition, RAP publishers, Bangkok, Thailand, pp19-21

Jayasekera RJ, Freitas MC (2005) Concentration Levels of Major and Trace Elements in Rice from Sri Lanka as Determined by the k 0 Standardization Method. Biol Trace Elem Res 103:83–96

Jayasumana C, Gunatilake S, Senanayake P (2014) Glyphosate, hard water and nephrotoxic metals: Are they the culprits behind the epidemic of chronic kidney disease of unknown etiology in Sri Lanka? Int J Environ Res Public Health 11:2125–2147

Jayasumana MACS, Paranagama PA, Amarasinghe MD et al (2013) Possible link of Chronic arsenic toxicity with Chronic Kidney Disease of unknown etiology in Sri Lanka. J Nat Sci Res 3:64–73

Jayatilake N, Mendis S, Maheepala P, Mehta FR (2013) Chronic kidney disease of uncertain aetiology: prevalence and causative factors in a developing country. BMC Nephrol 14:180

Kapaj S, Peterson H, Liber K, Bhattacharya P (2006) Human health effects from chronic arsenic poisoning–a review. J Environ Sci Health A Tox Hazard Subst Environ Eng 41:2399–2428

Meharg AA (2004) Arsenic in rice–understanding a new disaster for South-East Asia. Trends Plant Sci 9:415–417

Meharg AA, Hartley-Whitaker J (2002) Arsenic uptake and metabolism in arsenic resistant and nonresistant plant species. New Phytol 154:29–43

Meharg AA, Zhao FJ (2012) Arsenic and Rice. Springer Science & Business Media, London

Mukhopadhyay R, Bhattacharjee H, Rosen BP (2014) Aquaglyceroporins: generalized metalloid channels. Biochim Biophys Acta 1840:1583–91

Narukawa T, Matsumoto E, Nishimura T, Hioki A (2014) Determination of Sixteen Elements and Arsenic Species in Brown, Polished and Milled Rice. Anal Sci 30:245–250

Potera C (2007) Food Safety: U.S. Rice Serves Up Arsenic. Environ Health Perspect 115:A296

Premakumara GAS, Abeysekera WKSM, Ratnasooriya WD et al (2013) Antioxidant, anti-amylase and anti-glycation potential of brans of some Sri Lankan traditional and improved rice (Oryza sativa L.) varieties. J Cereal Sci 58:451–456

Raab A, Baskaran C, Feldmann J et al (2009) Cooking rice in a high water to rice ratio reduces inorganic arsenic content. J Env Monit 11(1):41–44

Rahman MA, Hasegawa H (2011) High levels of inorganic arsenic in rice in areas where arsenic-contaminated water is used for irrigation and cooking. Sci Total Environ 409:4645–4655

Ruangwises S, Saipan P, Tengjaroenkul B et al (2012) Total and Inorganic Arsenic in Rice and Rice Bran Purchased in Thailand. J Food Protection 75(4):771–774

Srivastava S, Suprasanna P, D'Souza SF (2012) Mechanisms of arsenic tolerance and detoxification in plants and their application in transgenic technology: a critical appraisal. Int J Phytoremediation 14:506–17

Tchounwou P, Patlolla A, Centeno J (2003) Carcinogenic and Systemic Health Effects Associated with Arsenic Exposure - A Critical Review. Toxicol Pathol 31:575–588

Tsuji JS, Yost LJ, Barraj LM et al (2007) Use of background inorganic arsenic exposures to provide perspective on risk assessment results. Regul Toxicol Pharmacol 48:59–68

Yost LJ, Schoof RA, Aucoin R (1998) Intake of Inorganic Arsenic in the North American Diet. Hum Ecol Risk Assess An Int J 4:137–152

Zhu YG, Williams PN, Meharg AA (2008) Exposure to inorganic arsenic from rice: A global health issue? Environ Pollut 154:169–171

9

Assessment of contamination of the beach clam *Tivela mactroides*: implications for food safety of a recreational and subsistence marine resource in Caraguatatuba Bay, Brazil

Márcia Regina Denadai[1,2*], Daniela Franco Carvalho Jacobucci[3], Isabella Fontana[4], Satie Taniguchi[2] and Alexander Turra[2]

Abstract

Background: The clam *Tivela mactroides* is an important sandy-beach resource along the western Atlantic coast, and is widely harvested by both tourists and residents for recreational, subsistence and/or economic purposes. These clams are intensively exploited in Caraguatatuba Bay on the southeastern Brazilian coast. Similarly to most coastal areas around the world, this bay is subject to a variety of environmental threats derived from human occupation (e.g., sewage) and economic activities (e.g., oil spills). Considering the history of changes in this area and current plans for development, environmental pressures are expected to increase. This prospect raises concerns regarding food safety of members of the public, including clam harvesters, who consume local seafood. In order to provide baseline information to compare with future situations, this study analyzed the contamination of clam meat by microorganisms (fecal coliforms, *Salmonella* sp., *Vibrio cholerae*, and *Staphylococcus aureus*) and Polycyclic Aromatic Hydrocarbons (PAHs).

Results: Preliminary evaluations revealed microorganism contamination levels above the maximum limits allowed under Brazilian legislation; with higher levels in the central portion of the bay. Temporal evaluations at three sampling points in this area revealed year-round contamination by all microorganisms, i.e., a continuous risk for clam consumers. Although the effectiveness of thermal processing used by consumers could not be formally tested in this study, it has the potential to reduce the contamination by fecal coliforms, *Salmonella* sp., and *S. aureus* to safe levels, as demonstrated in the two samples analyzed. However, although *S. aureus* can be totally eliminated, its heat-tolerant toxins may still affect consumers. Concentrations of individual compounds (congeners) and total PAHs were recorded, indicating contamination derived from oil spills.

Conclusions: The results raise concerns regarding traditional small-scale fisheries, which can be threatened by the intensification of human activities in the coastal region, thus requiring continuous monitoring of the quality of seafoods, in addition to effective communication of the risks to consumers, and efficient measures to reduce both sewage and industrial pollution.

Keywords: Clams; Coliforms; *Salmonella* sp; *Vibrio cholerae*; *Staphylococcus aureus*; Hydrocarbons; Caraguatatuba Bay

* Correspondence: marciard@gmail.com
[1]Departamento de Biologia Animal, Instituto de Biologia, Universidade Estadual de Campinas, Campinas, SP 13083-862, Brazil
[2]Instituto Oceanográfico, Universidade de São Paulo, São Paulo, SP 05508-900, Brazil
Full list of author information is available at the end of the article

Background

Seafood is an important source of protein, but the stocks and safety of fish and shellfish are being impacted by overfishing and marine pollution. Coastal areas are subject to a variety of conflicting interests and activities (Cicin-Sain and Knecht 1998), which reduce water quality and food safety. Sewage outfalls and nutrient enrichment as well as oil spills (Islam and Tanaka 2004; Diaz and Rosemberg 2008) are among the factors responsible for worldwide impacts in the sea (Halpern et al. 2008) and a general reduction in ocean health (Halpern et al. 2012). This situation is expected to worsen in developing countries, where monitoring and control policies are insufficient to mitigate coastal pollution (Wingqvist et al. 2012). This is a challenge even in developed countries such as the USA, where 6.1% (589,310 outbreaks) of the 9,638,301 foodborne outbreaks, between 1998 and 2008, were caused by consumption of contaminated aquatic animals (Painter et al. 2013), with bacteria identified as the etiological agent in 24% of the cases.

Food safety emerged as one of the central themes of the United Nations Conference on Sustainable Development, held in 2012 in Rio de Janeiro (UNCSD United Nations Conference on Sustainable Development 2012), in addition to ocean sustainability and poverty reduction. These themes interact in the context of environmental degradation of coastal zones, where impaired food safety may increase poverty and compromise human well-being. This is especially applicable to developing countries (Andrew et al. 2007) where traditional small-scale fisheries comprise a fragile socioecological system, prone to collapse under environmental degradation.

One type of small-scale fishery that might be especially endangered by coastal impacts is beach (or estuarine) clam harvesting (McLachlan et al. 1996; Castilla and Defeo 2001; Defeo et al. 2009). This fishery is affected not only by habitat alterations that reduce clam stocks, such as beach erosion, but also by the risks to humans consuming clams from polluted areas. According to Martinez and Oliveira (2010), counts of coliform bacteria are always higher in mollusks than in the water, due to their filtering habit and bioaccumulation, which increases the risks of consuming them. Consumption of mollusks was responsible for 3% of the cases of all foodborne disease recorded in the United States from 1998 to 2008 (Painter et al. 2013). Mollusks are related with 42.5% of seafood-associated outbreaks (Iwamoto et al. 2010), raising concerns for areas where mollusk consumption is intense.

According to Gelli et al. (1979), microbiological and chemical analyses of bivalve mollusks can indicate the microbiota and the contaminants that are present in the marine environment. This is the background for the worldwide Mussel Watch Program (Kimbrough et al.

2008), which uses this kind of information as a sentinel for public health, to control foodborne infections (Dias et al. 2003). Accordingly, concerns have increased and consequently new studies are being conducted on monitoring and assessment of microbiological contamination in marine mollusks, especially cultivated species (Pinheiro Jr 2000; Barardi et al. 2001; Garcia 2005), leaving a gap in information on the natural stocks.

The risks involved in beach-clam harvesting depend on the amount of clams collected and consumed, but also on their contamination levels, which are expected to increase in developing countries due to intense coastal development. One example of this situation is the southeastern coast of the states of São Paulo and Rio de Janeiro in Brazil, where most marine extraction of oil and natural gas in the country takes place. Between 1978 and 2004, 268 environmental accidents involving ships and oil pipelines were recorded near the São Sebastião Channel on the northern coast of São Paulo (São Paulo 2004), where more than half of the country's oil is transferred to the continent. The prospective negative environmental impacts of exploitation of the pre-salt oil reserves (Oxford Analytica 2010) may impact environmental quality in this region, especially due to its complex geomorphology and relatively low hydrodynamics, which result in areas that are more likely to retain the oil (Gherardi and Cabral 2007).

In addition to the pre-salt oil exploitation, several projects are being planned or implemented on the northern coast of São Paulo that will have individual and cumulative impacts (Legaspe 2012; Teixeira 2013). Specifically, the prospect for increased population density in this area is expected to put pressure on the existing infrastructure for collection and treatment of domestic sewage. This is a significant bottleneck in the region, since the present facilities are inadequate to the demand, and cause contamination that makes bathing at most beaches inadvisable (São Paulo 2005a). The number of acute diarrhea cases in the region increased from 1,007 to 8,046 between 1998 and 2003 (São Paulo 2005b). The local bays, which have low hydrodynamics, receive large amounts of fresh water and urban effluents from rivers that flow through the urban zones, as in Caraguatatuba Bay, where there is traditional and intense harvesting of a bivalve mollusk, the clam Tivela mactroides (Born 1778). The "berbigão", as it is locally named, can be found buried in the surface sediment layers in the intertidal and shallow subtidal zones (Denadai et al. 2005) and is an important resource for recreation, subsistence and economic input for tourists and residents (Denadai et al. in press). In 2005, an estimated 27 tons of clams were consumed in Caraguatatuba Bay by around 590 families or 2,183 individuals (Denadai et al. in press).

This study addressed the food safety of the *T. mactroides* fishery in Caraguatatuba Bay, as a baseline to understand the eventual impacts of future coastal changes. We focused on the two most important threats, microbiological contamination due to sewage outfalls, and chemical toxicity due to potential oil spills. The spatial and temporal occurrence of pathogenic bacteria was evaluated, as well as the capacity of thermal treatment to reduce contamination to safe levels before consumption. We selected the bacteria that are traditionally considered indicative of fecal contamination (fecal coliforms; Huss 1997), that are historically recorded or have high pathogenicity in estuarine environments (*Staphylococcus aureus* and *Salmonella* sp.; Huss 1997), and that are typical of marine environments (the indigenous *Vibrio cholerae*; Huss 1997). Polycyclic Aromatic Hydrocarbons (PAHs) are important organic pollutants, mostly from anthropogenic (pyrolytic or petrogenic) activities (Woodhead *et al.* 1999), and merit special attention due to their potential negative effects and continuous input into the environment (Yunker *et al.* 2002). As PAHs may serve as indicators of oil spills in the region, and in order to improve understanding of their potential risks to humans, their concentrations were evaluated in different areas in Caraguatatuba Bay.

Methods

Study area and sampling procedure

The study was conducted in Caraguatatuba Bay on the northern coast of São Paulo state, Brazil, from 2003 through 2005. This bay is bounded by a 16 km-long sandy beach, which is crossed by four rivers that collect and transport continental effluents into the bay (Figure 1).

Initially, three collection expeditions were conducted, on March 30, 2003 and January 10 and February 26, 2004, to screen for the presence of the selected bacteria in clam meat. Six samples were collected in each period at different sites along the shore, in order to cover the clam harvesting area (Figure 1; points 1–6).

Subsequently to this pilot sampling, year-round samples were collected to determine the temporal variation in levels of bacterial contamination. Three new sampling sites (A, B, and C) were defined in the area where both clam harvesting intensity (Denadai et al. in press) and the preliminary results on contamination were highest (Figure 1). At each site, samples of beach clams were collected on April 22, May 19 and September 2, 2004, and December 2, 2005. The contamination levels for fecal coliforms and *S. aureus* were compared along the

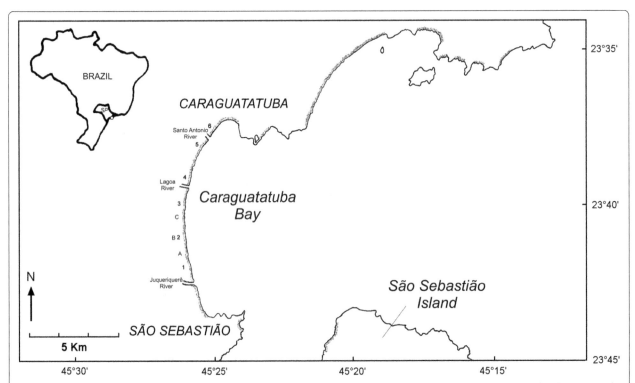

Figure 1 Map of Caraguatatuba Bay showing the sampling sites. Site 1 - 23°41'43.5"S 45°25'43.7"W; 2 - 23°40'53.5"S 45°25'49.9"W; 3 - 23°39'57.9"S 45°25'46.2"W; 4 - 23°39'12.9"S 45°25'36.0"W; 5 - 23°38'15.5"S 45°25'11.8"W; 6 - 23°37'48.4"S 45°24'52.9"W; Sample A - 23°41'26.8"S 45°25'47.4"W; B - 23°40'51.7"S 45°25'51.2"W; C - 23°40'04.6"S 45°25'49.0"W.

time series with a one-factor analysis of variance (ANOVA) followed by a Student–Newman–Keuls (SNK) test for pairwise comparisons (Underwood 1997). The results for *V. cholerae* and *Salmonella* sp. were not analyzed quantitatively, since the analysis is based on their detection (presence/absence). The mere presence of these bacteria makes seafood unsafe for human consumption, under current legislation (see below).

At the same time, we analyzed the potential effect of thermal processing in reducing the microbiological contamination. The samples from sites 2 and 5, collected on February 26, 2004, were analyzed before and after the heating procedure, which consisted basically of the caiçaras' (traditional coastal residents of southeastern Brazil) practice of cooking the clams in boiling water for 10 minutes (Denadai et al. in press).

Each sample for microbial-contamination analysis consisted of 25 g of clam meat, which is equivalent to eight individual clams approximately 30 mm in shell length. The clams were collected during low tide and immediately refrigerated (<10°C for 24 h). In the laboratory, the shells were cleaned with 70% ethanol in a laminar-flow hood and their soft tissues were extracted with a sterile spatula. This pooled sample size was based on the regulation of the Brazilian Health Surveillance Agency (ANVISA) for detection of *Salmonella* sp. (Brasil, Agência Nacional de Vigilância Sanitária 2001). The only specified regulation for raw samples established at the time of the sample collection was for *Salmonella* sp., requiring 25 g of clam meat. For this reason, we used 25-g samples for the analyses of all other contaminants.

We followed the methods of Downes and Ito (2001) for all microbiological analyses. The samples of clam meat were aseptically taken and homogenized with 225 mL of sterile 0.2% peptone water using a stomacher for 2 minutes. Samples were used for fecal coliforms analyses by Most Probable Number (MPN/g) method specified in ISO 21528–1:2004 (Thermo Fisher 2013). For the isolation of *Vibrio cholerae* we utilized TCBS Agar (HiMedia Laboratories, Mumbai, India) that was developed by Kobayashi *et al.* (1963). TCBS Agar is also recommended by the American Public Health Association (APHA) for the selective isolation of *V. cholerae* that shows yellow color colonies on the TCBS agar surface. For the isolation of *Salmonella* species we used HiCrome *Salmonella* Agar (Sigma-Aldrich, St. Louis, MO, USA), a selective medium used for simultaneous detection of *Escherichia coli* and *Salmonella* from food and water. *Escherichia coli* and *Salmonella* are distinguishable due to the colony characteristics since *Salmonella* give light purple colonies with halo. For the isolation of *Staphylococcus aureus* we used Baird-Parker Agar Base (BD Difco, Franklin Lakes, NJ, USA) with Egg Yolk Tellurite Enrichment in the preparation for selective isolation and enumeration (CFU/g; Colony-Forming Units) of coagulase-positive staphylococci from food. The typical colonies of *S. aureus* are black, shiny, convex and surrounded by clear zones of approximately 2–5 mm.

Concomitantly with the microbiological approach, we analyzed contamination by PAHs. For this baseline characterization, two samples of 200 g each of beach clams were collected on March 30, 2004. The first sample consisted of specimens from the northern part of the bay (sites 2 and 3) and the second sample of specimens from the southern part (sites 4 and 5) (Figure 1). The analyses were performed by CQA (Centro de Qualidade Analítica Ltda., Campinas, SP, Brazil) using methods 3550B and 8310 as recommended by the United States Environmental Protection Agency (EPA United States Environmental Protection Agency 1986, 1996). Clam tissues were dried and homogenized with sodium sulfate and extracted three times using an ultrasonic disruptor with a mixture of acetone/methylene chloride (1:1, v/v). The extract was purified in a silica gel chromatographic column, and PAHs were quantitatively analyzed by high-performance liquid chromatography (HPLC) coupled with fluorescence detection and excitation at 280 nm and emission at 389 nm. A reverse-phase column HC-ODS Sil-X, 5-micron particle size diameter, in a 250-mm × 2.6-mm I.D. stainless steel column was used in an isocratic elution for 5 min using acetonitrile/water (4:6)(v/v), then a linear gradient elution to 100% acetonitrile over 25 min at 0.5 mL/min flow rate. Certified PAH standards at five different concentrations were used to construct the analytical curve.

Results and discussion

The pilot study of microbiological contamination on March 30, 2003 and January 10 and February 26, 2004 showed that all the bacterial groups analyzed were present in the clam tissue (Table 1).

According to Brazilian regulations (Brasil, Agência Nacional de Vigilância Sanitária 2001), In order to be considered safe for consumption, clam tissues must contain no *Salmonella* sp. A single sample from each site was sufficient to detect *Salmonella* sp. in the samples collected on February 26, 2004. Its absence from the two samples collected in 2003 does not mean that the area was safe for shellfish harvesting at that time, since, according to Brasil, Agência Nacional de Vigilância Sanitária (2001) and FAO/WHO Food and Agriculture Organization of United Nations/World Health Organization (2008), five samples from each site must be negative for an area to be considered free of *Salmonella* sp.

Brazilian legislation establishes the maximum acceptable levels of fecal coliforms as follows: for cooked bivalve mollusks, spiced or not, factory-chilled or frozen, the ANVISA established that for five samples of 200 g of

Table 1 Microbiological contamination of meat of the clam *Tivela mactroides* sampled in 2003–2004 along the shore of Caraguatatuba Bay, southeastern Brazil

Agents	Unit	Ref. val. Brasil, Agência Nacional de Vigilância Sanitária (2001)	Ref. val. FAO/WHO Food and Agriculture Organization of United Nations/ World Health Organization (2008)	March 30, 2003 Samples						January 10, 2004 Samples						February 26, 2004 Samples					
				1	2	3	4	5	6	1	2	3	4	5	6	1	2	3	4	5	6
Fecal coliforms	MPN/g	50	700	920	170	110	8	13	8	460	14	2400	110	43	11	43	≥2400	1100	93	9	460
Staphylococcus aureus	CFU/g	1000		–	–	1700	3000	–	–	–	–	–	–	–	–	7000	12000	57000	3500	4000	5000
Salmonella sp.		Abs.	·	–	–	Abs.	Abs.	–	–	–	Pres.	Pres.	–	Pres.	Pres.	Pres.	Pres.	Pres.	Pres.	Pres.	Pres.
Vibrio cholerae		Abs.		–	–	–	–	–	–	–	Pres.	Pres.	Pres.	Pres.	Pres.	Pres.	Pres.	Pres.	Pres.	Pres.	Pres.

Ref. Val. = maximum reference limits for human consumption, MPN = most probable number, CFU = colony-forming units, Abs. = absence, Pres. = presence, – = No data.

the edible part, none of the samples may contain more than 50 MPN/g coliforms at 45°C and only two samples may contain between 10 and 50 MPN/g (Brasil, Agência Nacional de Vigilância Sanitária 2001). In 2008, after our sampling and processing had been completed, the FAO/WHO Commission established a regulation in the CODEX Alimentarius for live or raw bivalve mollusks in which, for five 100 g samples of edible parts, none may contain more than 700 *Escherichia coli* and no more than one sample may contain between 230 and 700 *E. coli*. However, the maximum acceptable levels of the CODEX Alimentarius were defined specifically for *E. coli* because it is a broader indicator of fecal contamination. Because our study evaluated fecal coliforms as a whole (not specifically *E. coli*) and also because we used 25 g of fresh meat, instead of the non-fresh 100/200 g mentioned in the previous resolutions, we decided to present as a reference limit for interpreting the results for fecal coliforms, both the values of Brasil, Agência Nacional de Vigilância Sanitária (2001) and also the values of the CODEX Alimentarius (FAO/WHO Food and Agriculture Organization of United Nations/World Health Organization 2008). Even though our pooled sample size was smaller than the ANVISA recommendation of 200 g for fecal-coliform analysis, the samples with 25 g were sufficient to detect a high level of contamination. On March 30, 2003, fecal coliform counts above the reference limits were recorded in three (according to Brasil, Agência Nacional de Vigilância Sanitária 2001) and one (according to FAO/WHO Food and Agriculture Organization of United Nations/World Health Organization 2008) of six samples. On January 10, 2004, in the summer vacation period, fecal coliform counts above the reference limits were recorded in three (Brasil, Agência Nacional de Vigilância Sanitária 2001) and one (FAO/WHO Food and Agriculture Organization of United Nations/World Health Organization 2008) of six samples. On February 26, 2004, just after the summer vacation period of 2004, fecal coliforms showed levels higher than reference values in four (Brasil, Agência Nacional de Vigilância Sanitária 2001) and two (FAO/WHO Food and Agriculture Organization of United Nations/World Health Organization 2008) of the six sites.

For raw, chilled, or frozen bivalve mollusks that will not be consumed raw, the maximum acceptable level of *Staphylococcus*-positive coagulase (= *S. aureus*) is 10^3 CFU/g (Brasil, Agência Nacional de Vigilância Sanitária 2001, FAO/WHO Food and Agriculture Organization of United Nations/World Health Organization 2012). In March 30, 2003, counts of *S. aureus* above the limits were recorded in each of two samples analyzed, and in February 26, 2004, *S. aureus* occurred above the limits at all six sites.

For *V. cholerae*, no reference values were found in the current legislation, so we considered absence as an acceptable quality indicator. *V. cholerae* was recorded in all samples collected in January 10, 2004 and February 26, 2004.

The most-contaminated sites in all periods (1 to 4) were those that were most used by clammers (Denadai et al. in press). Therefore, a new and more-detailed sampling strategy was implemented in the southern part of the study area, in four different periods year-round, with three samples (sites) per period.

In this new sampling series, *V. cholerae* and *Salmonella* sp. were present in all 12 samples analyzed (Table 2). The level of contamination with fecal coliforms varied among sampling periods (ANOVA, F = 524.46, df = 3, p < 0.001), with the highest values recorded on 22 April, 2004 (SNK, p < 0.001 for all comparisons) (Table 2), during a long national holiday, when the population increased significantly along the coast. Fecal-coliform contamination levels exceeded the reference limits in six (Brasil, Agência Nacional de Vigilância Sanitária 2001) and three (FAO/WHO Food and Agriculture Organization of United Nations/World Health Organization 2008) of the 12 samples. *S. aureus* did not show significant temporal variation (ANOVA, F = 2.75, df = 3, p = 0.113), but the highest individual values (3,900 CFU.g^{-1}) were found in April 2004, and the lowest (300 CFU.g^{-1}) in February 2005. Contamination of *S. aureus* above the limits was found in six of the 12 samples, and only the samples taken on February 16, 2005 showed no value above the limits.

Although no published records of beach-clam contamination in Brazil were found, contamination with pathogenic bacteria has been an important factor in Brazilian mariculture. Studies with cultivated mollusks have revealed considerable levels of contamination. Pereira *et al.* (2006) detected the frequent presence of fecal coliforms in the cultivated oyster *Crassostrea gigas* sold in markets in the state of Santa Catarina. *E. coli* was present in 9% of the cultivated samples and in 35.5% of the market samples, and although Pereira *et al.* (2006) did not detect *Salmonella* sp. or *Vibrio* sp., *S. aureus* was present in only one sample, with 80 CFU/g, while the remaining samples contained <10 CFU/g. On the Ceará coast, Vieira *et al.* (2008) found that contamination of the cultivation water by fecal coliforms remained within the allowed limits. Only one of 15 samples showed high levels of total coliforms (2,800 MPN/100 mL) and thermotolerant coliforms (3,500 MPN/100 mL). These results indicate contamination of the beach clams from Caraguatatuba Bay, since the frequency and the absolute scores for fecal-coliform contamination were higher than those found by other investigators in cultivated mollusks (Pereira *et al.*

Table 2 Microbiological contamination of meat of the clam *Tivela mactroides* sampled in 2004–2005 along the shore of Caraguatatuba Bay, southeastern Brazil

Agents	Unit	Ref. val.		April 22, 2004 Sampling sites			May 19, 2004 Sampling sites			September 2, 2004 Sampling sites			February 16, 2005 Sampling sites		
		Brasil, Agência Nacional de Vigilância Sanitária (2001)	FAO/WHO Food and Agriculture Organization of United Nations/World Health Organization (2008)	A	B	C	A	B	C	A	B	C	A	B	C
Fecal coliforms	MPN/g	50	700	≥2400	≥2400	≥2400	210	240	15	9	4	9	240	39	9
Staphylococcus aureus	CFU/g	1000		3500	610	3900	2400	1500	500	1400	1200	800	300	300	300
Salmonella sp.		Abs.		Pres.	Pres.	Pres.	Pres.	Pres.	Pres.	Pres.	Pres.	Pres.	Pres.	Pres.	Pres.
Vibrio cholerae		Abs.		Pres.	Pres.	Pres.	Pres.	Pres.	Pres.	Pres.	Pres.	Pres.	Pres.	Pres.	Pres.

Ref. Val. = maximum reference limits for human consumption, MPN = most probable number, CFU = colony-forming units, Abs. = absence, Pres. = presence.

Table 3 Microbiological contamination of meat from the clam *Tivela mactroides*, before and after thermal processing, sampled in Caraguatatuba Bay, southeastern Brazil

Agents	Unit	Ref. val.		Thermal processing			
		Brasil, Agência Nacional de Vigilância Sanitária (2001)	FAO/WHO Food and Agriculture Organization of United Nations/World Health Organization (2008)	Site 2		Site 5	
				Before	After	Before	After
Fecal coliforms	MPN/g	50	700	≥2400	<3	9	9
Staphylococcus aureus	CFU/g	1000		12000	Abs.	4000	Abs.
Salmonella sp.		Abs.		Pres.	Abs.	Pres.	Abs.
Vibrio cholerae		Abs.		Pres.	Abs.	Pres.	Pres.

Ref. Val. = maximum reference limits for human consumption, MPN = most probable number, CFU = colony-forming units, Abs. = absence, Pres. = presence.

2006; Vieira *et al.* 2008). We consider that the different factors such as the physical structure of the embayment, with its low hydrodynamics, and the large and diffuse sewage runoff in the region are responsible for these findings.

According to FAO/WHO Food and Agriculture Organization of United Nations/World Health Organization (2008), environmental parameters such as temperature and salinity are not predictive for *Salmonella* sp. contamination in areas of mollusk harvesting, and data that could quantify the probability of growth of this bacterium in mollusk tissues after harvesting are not available. A relationship exists between the fecal coliform concentration in the tissues and the probability of *Salmonella* sp. presence, but may vary according to the location.

Brazilian legislation mandates, for the natural or intensive farming of species destined for human consumption and that will be consumed raw, that the mean concentration of fecal coliforms in the water must not exceed 14 MPN/100 mL, and 10% of the samples must not exceed 43 MPN/100 mL (CONAMA Resolution No. 20/1986) (Brasil, Conselho Nacional de Meio Ambiente 1986). The beaches of Caraguatatuba Bay, monitored by the Companhia de Tecnologia de Saneamento Ambiental (CETESB), had a mean concentration of 81.8 MPN/100 mL during the study period (Paulo 2005a). Only 24% of the 260 water samples analyzed in 2004 had fecal coliform levels below 14 MPN/100 mL, and 48.5% had levels above 43 MPN/100 mL (Paulo 2005a). These data agree with the bacterial levels found in the beach-clam meat, and indicate that the sanitation conditions of the bay waters represent a high risk for harvesters and consumers of clams.

The most contaminated samples of clam meat were from the central part of the bay, where clamming is most common (Paulo 2005a, Denadai et al. in press). The data suggest an effect of summer holiday periods, when the increase in population may overload the inadequate sewage system. As rain and thus continental drainage (including non-point pollution sources) may increase the input of microbial contaminants, and as tourism is higher in summer months, which is the rainy season in southeastern Brazil, this period is considered the most risky, when clam collecting and consumption should be avoided and communication of this potential hazard to the public should be intensified.

We also applied the thermal treatment commonly used by locals (10 min in boiling water) to two of the samples. Even though this procedure was not conducted to formally evaluate the thermal treatment effectiveness, the results demonstrated that this cooking time totally eliminated *S. aureus* and *Salmonella* sp. and reduced fecal coliforms to acceptable levels in both samples (Brasil, Agência Nacional de Vigilância Sanitária 2001;

FAO/WHO Food and Agriculture Organization of United Nations/World Health Organization 2008). However, this cooking method did not eliminate *V. cholerae* in one of the samples (Table 3). This indicates that even though *V. cholerae* is very sensitive to high temperatures (D_{50} of 0.30 min at 71°C) (Shultz *et al.* 1984), the simple boiling procedure may not suffice to totally eliminate the microorganism. Makukutu and Guthrie (1986) found that *V. cholerae* can survive and remain viable after one hour exposed to 60°C.

The incomplete elimination of fecal coliforms can be explained by their known high heat tolerance (D_{50} of 45 min at 60°C) (Padhye and Doyle 1992). *Salmonella* sp. and *S. aureus* are more sensitive to heat, with D_{50} of 10 min at 60°C (Thomas *et al.* 1966) and 3 min at 55°C (Halpin-Dohnalek and Marth 1989). Even though *S. aureus* was completely eliminated, the toxins produced by this species are very heat-resistant (D_{50} of 68.5 min at 98.9°C) (Read and Bradshaw 1966) and might not be completely eliminated by cooking, thus posing an additional threat to consumers.

Brazilian legislation has no established criteria for the maximum PAHs allowed in foods. We used the limit from European Regulation 1881/2006/EC (ER) of 10 µg kg^{-1} dw in bivalve mollusks for benzo[a]pyrene, used as a reference for toxicity of individual PAHs because it is one of the most toxic compounds in this group (Wenzl *et al.* 2006). Only low-molecular-weight compounds (2 or 3 fused aromatic rings), such as naphthalene, acenaphthylene, anthracene, fluorene and phenanthrene were found (Table 4). They are more

Table 4 Concentrations of PAHs in meat from the clam *Tivela mactroides* sampled in Caraguatatuba Bay, southeastern Brazil

Compounds	South area (µg/kg)	North area (µg/kg)
Naphthalene	1.90	1.80
Acenaphthylene	0.20	0.20
Fluorene	0.40	<0.01
Anthracene	0.90	<0.01
Pyrene	<0.01	<0.01
Benz[a]anthracene	<0.01	<0.01
Chrysene	<0.01	<0.01
Benzo[k]fluoranthene	<0.01	<0.01
Benzo[a]pyrene	<0.005	<0.005
Dibenzo[a,h]anthracene	<0.01	<0.01
Benzo[g,h,i]perylene	<0.01	<0.01
Indeno[1,2,3]pyrene	<0.01	<0.01
Phenanthrene	1.00	1.40
Fluoranthene	<0.01	<0.01
Total PAH	4.40	3.40

soluble in water, and their presence, even in relatively low concentrations, can be indicative of petroleum-derivative inputs at the two sampling sites (Villeneuve *et al.* 1999). Based on the limit in the European Regulation (Wenzl *et al.* 2006), no compounds in this study, even the total PAHs, showed concentrations above 10 µg kg^{-1} dw, and thus would not put consumers at risk. Particularly in view of the single sampling point and also the small sample size, the presence of oil compounds in the clam meat indicates the presence of pollution in the region, which is threatened by the presence of petroleum industry facilities and activities. As these compounds are bioaccumulated, it is important to analyze their levels periodically in clams over 25 mm, which require seven to ten months to reach this size, depending on the year (Turra *et al.* 2014); as well as after oil spills in the region, to assess the possible risks to consumers.

Conclusions

With the sample size and strategy employed here it was not possible to estimate the true prevalence of contamination in Caraguatatuba Bay, but it was possible to detect the presence of the pathogenic agents and contaminants investigated. The results provide evidence that in the 2003–2005 period, the health conditions of Caraguatatuba Bay were unsatisfactory, causing bacterial contamination of the clam meat, including at higher levels than allowed under Brazilian health regulations and possibly related to the sites that are most used by clammers. Complementarily, recent reports on bathing conditions in Caraguatatuba Bay have shown no improvement in water quality since the study period.

Based on this case study of *T. mactroides*, socioecological systems based on beach-clam fisheries, which have recreational, subsistence and economic importance, may be considered at high risk worldwide. Education about public health and adequate sewage treatment must be provided by the public sector, to guarantee safer consumption and environmental health. Thus, in addition to the education measures required to adequately inform the general public about the risks of consuming clams and the best way to prepare them in order to reduce or eliminate possible infectious agents, additional efforts are needed to reduce the risks of consuming non-regulated seafood. Extensive expansion of the sewage system is urgent in order to reduce the runoff of household waste, as well as to improve sewage treatment for safe disposal in the sea. This requires attention from the public sector and investment in sanitation facilities and equipment, mainly due to the recent Caraguatatuba population increase (19% between 2004 and 2013; São Paulo 2013) and the expected future increase due to the different industrial projects that are being implemented in the region. During this period (2004–2013) the sanitation conditions of the beaches and rivers entering Caraguatatuba Bay oscillated, with recent worsening of water quality (Paulo 2013); from 2011 to 2013, the water quality was within acceptable limits ≤ 25% of the time. A report by CETESB on thermotolerant coliforms, *E. coli*, and enterococcus, indicated that the beaches in Caraguatatuba Bay were classified as acceptable (Pan Brasil, Palmeiras, and Porto Novo), poor (Centro) or very poor (Indaiá). Considering that the criteria used to classify water quality for bathing are less stringent than those used to classify zones for food production, the results of this study are probably consistent with the current situation in Caraguatatuba Bay. In addition to the various threats to the habitats and stocks of the clams themselves, increasing concern regarding food safety may lead to interdictions on clamming, which is a very important economic activity for low-income fishing communities in developing countries.

The development of a regulated surveillance program based on continuous monitoring of both biological and chemical contaminants in the water (continuously evaluated by CETESB) and in *T. mactroides* is strongly recommended to detect possible risks to humans who consume these clams. Beach-clam fisheries appear to be an important socioecological indicator of coastal resilience and health, and should be incorporated into future assessments of ocean quality.

Abbreviations
ANOVA: Analysis of variance; ANVISA: Agência Nacional de Vigilância Sanitária; CETESB: Companhia de Tecnologia de Saneamento Ambiental; CFU: Colony-forming units; MPN: Most probable number; PAHs: Polycyclic Aromatic Hydrocarbons; SNK: Student–Newman–Keuls.

Competing interests
The authors declare that they have no competing interests.

Authors' contributions
MRD planed the sample and analytical strategies, carried out the samples, analyzed the data and revised the manuscript in all the phases. DFCJ carried out the microbiological analysis and revised the manuscript. IF prepared the final version of the manuscript and incorporated the reviewers suggestions. ST prepared the text on HPA contamination and revised the manuscript. AT planed the sample and analytical strategies, carried out the samples, analyzed the data and revised the manuscript in all the phases. All authors read and approved the final manuscript.

Acknowledgments
We thank FAPESP (Proc. No. 01/06955-5), PADI Foundation, and PROJECT AWARE Foundation for their financial support of this project. We also thank the trainees Carla Guinart Marques, Fernanda Aparecida Ribeiro, Ana Lúcia Lima, Francini Migotto Cabral, Adriana Maciel Leituga Elias, Fernanda Navarro Junqueira Anadão, and Leilane Barbosa Ronqui for their assistance in field and laboratory activities. Our thanks to Dr. Janet W Reid, who revised the English text, and to two anonymous referees, whose comments significantly improved the article.

Author details
[1]Departamento de Biologia Animal, Instituto de Biologia, Universidade Estadual de Campinas, Campinas, SP 13083-862, Brazil. [2]Instituto Oceanográfico, Universidade de São Paulo, São Paulo, SP 05508-900, Brazil.

[3]Instituto de Ciências Biomédicas, Instituto de Biologia, Universidade Federal de Uberlândia, Uberlândia, MG 38400-902, Brazil. [4]Departamento de Medicina Veterinária Preventiva e Saúde Animal, Faculdade de Medicina Veterinária e Zootecnia, Universidade de São Paulo, São Paulo, SP 05508-270, Brazil.

References

Andrew NL, Béné C, Hall SJ, Allison EH, Heck S, Ratner BD (2007) Diagnosis and management of small-scale fisheries in developing countries. Fish 8:227–240

Barardi CRM, dos Santos CS, Simões CMO (2001) Ostras de qualidade em Santa Catarina. Ciência Hoje 29(172):70–73

Brasil, Agência Nacional de Vigilância Sanitária (2001) Resolução RDC n° 12, de 2 de janeiro de 2001. Aprova o Regulamento Técnico sobre padrões microbiológicos para alimentos. Diário Oficial da União, Brasília, DF

Brasil, Conselho Nacional de Meio Ambiente (1986) Resolução n° 20 de 18 de Junho de 1986. Diário Oficial da União, Brasília, DF

Castilla JC, Defeo O (2001) Latin American benthic shellfisheries: Emphasis on co-management and experimental practices. Rev Fish Biol Fish 11(1):1–30

Cicin-Sain B, Knecht RW (1998) Integrated Coastal and Ocean Management: Concepts and Practices. Island Press, Washington, DC, p 519

Defeo O, McLachlan A, Schoeman DS, Schlacher TA, Dugan J, Jones A, Lastra M, Scapini F (2009) Threats to sandy beach ecosystems: a review. Estuar Coast Shelf Sci 81:1–12

Denadai MR, Amaral ACZ, Turra A (2005) Along- and across-shore components of the spatial distribution of the clam Tivela mactroides (Born, 1778) (Bivalvia, Veneridae). J Nat Hist 39(36):3275–3295

Denadai MR, Pombo M, Bernadochi LC, Turra A (in press) Harvesting of the clam Tivela mactroides in southeastern Brazil: short- and long-term dynamics, with estimates of the amount of clams harvested and of consumers. Marine and Coastal Fisheries: 7(1). DOI: 10.1080/19425120.2015.1007183

Dias JFB, Mesquita EFM, Franco RM, Jesus EFO, Oliveira LAT (2003) Redução da carga bacteriana da ostra nativa [Crassostrea rhizophorae (Guilding, 1828)] in natura, resfriada e congelada, através da radiação gama. Higiene Alimentar 224:56

Diaz RJ, Rosemberg R (2008) Spreading dead zones and consequences for marine ecosystems. Science 321(5891):926–929

Downes FP, Ito K (eds) (2001) Compendium of Methods for the Microbiological Examination of Foods, 4th edn. American Public Health Association, Washington, DC

EPA United States Environmental Protection Agency (1986) Test Methods for Evaluating Solid Waste, Physical/Chemical Methods. Method 8310, vol 1, p 13. http://www.epa.gov/solidwaste/hazard/testmethods/sw846/pdfs/8310.pdf.

EPA United States Environmental Protection Agency (1996) Test Methods for Evaluating Solid Waste Physical/Chemical Methods, SW-846, Method 3550B: Ultrasonic Extraction, Revision 2, p 14, http://www.trincoll.edu/~henderso/textfi~1/416%20notes/3550b.pdf

FAO/WHO Food and Agriculture Organization of United Nations/ World Health Organization (2012) Salmonella in bivalve mollusks. CODEX Alimentarius Commission, Joint FAO/WHO Food Standards Programme, Codex Committee On Fish and Fishery Products, CX/FFP 12/32/2-Add.1, pp 1–6

FAO/WHO Food and Agriculture Organization of United Nations/World Health Organization (2008) Standard for Live and Raw Bivalve Molluscs. CODEX Alimentarius International Food Standards, CODEX STAN 292–2008, Rome, Italy. p 6

Garcia AN (2005) Contaminação microbiológica na área de cultivo de moluscos bivalves de Anchieta (Espírito Santo, Brasil). Graduation monograph, Oceanografia, Centro de Ciências Humanas e Naturais, Universidade Federal do Espírito Santo, Vitória, ES, Brazil. p 68

Gelli DS, Tachibana T, Sakuma H (1979) Ocorrência de Vibrio parahaemolyticus, Escherichia coli e de bactérias mesófilas em ostras. Rev Inst Adolfo Lutz 39(1):61–66

Gherardi DFM, Cabral AP (2007) Atlas de Sensibilidade Ambiental ao Óleo da Bacia Marítima de Santos. Secretaria de Mudanças Climáticas e Qualidade Ambiental. Ministério do Meio Ambiente, Brasília, p 124

Halpern BS, Walbridge S, Selkoe KA, Kappel CV, Micheli F, D'Agrosa C, Bruno JF, Casey KS, Ebert C, Fox HE, Fujita R, Heinemann D, Lenihan HS, Madin EMP, Perry MT, Selig ER, Spalding M, Steneck R, Watson R (2008) A global map of human impact on marine ecosystems. Science 319(5865):948–952

Halpern BS, Longo C, Hardy D, McLeod KL, Samhouri JF, Katona SK, Kleisner K, Lester SE, O'Leary J, Ranelletti M, Rosenberg AA, Scarborough C, Selig ER, Best BD, Brumbaugh DR, Chapin FS, Crowder LB, Daly KL, Doney SC, Elfes C,

Fogarty MJ, Gaines SD, Jacobsen KI, Karrer LB, Leslie HM, Neeley E, Pauly D, Polasky S, Ris B, St Martin K et al (2012) An index to assess the health and benefits of the global ocean. Nature 488:615–120

Halpin-Dohnalek MI, Marth EH (1989) Staphylococcus aureus: production of extracellular compounds and behavior in foods - a review. J Food Prot 52(4):267–282

Huss HH (1997) Garantia da Qualidade dos Produtos da Pesca. Documento Técnico Sobre as Pescas, 334. FAO, Rome, p 176

Islam MS, Tanaka M (2004) Impacts of pollution on coastal and marine ecosystems including coastal and marine fisheries and approach for management: a review and synthesis. Mar Pollut Bull 48(7–8):624–649

Iwamoto M, Ayers T, Mahon BE, Swerdlow DL (2010) Epidemiology of seafood-associated infections in the United States. Clin Microbiol Rev 23(2):399–411

Kimbrough KL, Johnson WE, Lauenstein GG, Christensen JD, Apeti DA (2008) An Assessment of Two Decades of Contaminant Monitoring in the Nation's Coastal Zone. NOAA Technical Memorandum NOS NCCOS 74, Silver Spring, MD, USA. p 105

Kobayashi T, Enomoto S, Sakazaki R, Kuwahara S (1963) A new selective medium for Vibrio group (modified Nakanishi's medium - TCBS agar). Nippon Saikingaku Zasshi 18(10,11):387–392

Legaspe LBC (2012) Os potenciais impactos cumulativos das grandes obras - novo corredor de exportação e exploração de hidrocarbonetos do campo mexilhão - no território da APA Marinha Litoral Norte (SP). M.SC. Dissertation, Instituto de Geociências e Ciências Exatas, Universidade Estadual Paulista, Rio Claro, SP, Brazil. p 111

Makukutu CA, Guthrie RK (1986) Behavior of Vibrio cholerae in hot foods. Appl Environ Microbiol 52(4):824–831

Martinez DI, Oliveira AJFC (2010) Fecal bacteria in Perna perna (Linnaeus, 1758) (Mollusca: Bivalvia) for biomonitoring coastal waters and food quality. Braz J Oceanography 58(3):29–35

McLachlan A, Dugan JE, Defeo O, Ansell AD, Hubbard DM, Jaramillo E, Penchaszadeh PE (1996) Beach clam fisheries. Oceanography Mar Biol An Annual Review 34:163–232

Oxford Analytica (2010) The impact of Pre-Salt: a long-term perspective. A report prepared for PETROBRAS, Oxford, UK, p 43

Padhye NV, Doyle MP (1992) Escherichia coli O157:H7: Epidemiology, pathogenesis, and methods for detection in foods. J Food Prot 55(7):555–565

Painter JA, Hoekstra RM, Ayers T, Tauxe RV, Braden CR, Angulo FJ, Griffin PM (2013) Attribution of foodborne illnesses, hospitalizations, and deaths to food commodities by using outbreak data, United States, 1998–2008. Emerg Infect Dis 19(3):407–415

Pereira MA, Nunes MM, Nuernberg L, Schulz D, Batista CRV (2006) Microbiological quality of oysters (Crassostrea gigas) produced and commercialized in the coastal region of Florianópolis. Braz J Microbiol 37:159–163

Pinheiro AA Jr (2000) Colimetria de águas marinhas e mexilhões (Perna perna LINNAEUS, 1758) em áreas de cultivo e extrativismo no município de Niterói, RJ. M.Sc. Dissertation, Faculdade de Medicina Veterinária, Universidade Federal Fluminense, Niteroi, RJ, Brazil. p 70

Read RB Jr, Bradshaw JG (1966) Staphylococcal enterotoxin B thermal inactivation in milk. J Dairy Sci 49:202–203

São Paulo (2004) Relatório de acidentes ambientais. Governo do Estado de São Paulo, Secretaria de Estado do Meio Ambiente. Companhia de Tecnologia de Saneamento Ambiental, São Paulo

São Paulo (2005a) Relatório de qualidade das águas litorâneas do estado de São Paulo: balneabilidade das praias 2004. Governo do Estado de São Paulo, Secretaria de Estado do Meio Ambiente. Companhia de Tecnologia de Saneamento Ambiental, São Paulo, p 183

São Paulo (2005b) Litoral norte. Secretaria do Meio Ambiente, Coordenadoria de Planejamento Ambiental. Estratégico e Educação Ambiental, São Paulo, p 112

São Paulo (2013) Relatório de qualidade das praias litorâneas no estado de São Paulo 2013. Governo do Estado de São Paulo, Secretaria de Estado do Meio Ambiente. Companhia de Tecnologia de Saneamento Ambiental, São Paulo, p 213

Shultz LM, Rutledge JE, Grodner RM, Biede SL (1984) Determination of the thermal death of Vibrio cholerae in blue crabs (Callinectes sapidus). J Food Prot 47(1):4–6

Teixeira LR (2013) Megaprojetos no Litoral Norte Paulista: O Papel dos Grandes Empreendimentos de Infraestrutura na Transformação Regional. Ph.D. Thesis, Universidade Estadual de Campinas, Instituto de Filosofia e Ciências Humanas, Campinas, SP, Brazil. p 274

Thermo Fisher Scientific (2013) Making food safer according to ISO methods. http://www.oxoid.com/pdf/iso-food-safety-brochure.pdf. Accessed 16 March 2015

Thomas CT, White JC, Longrée K (1966) Thermal resistance of *Salmonellae* and *Staphylococci* in foods. Appl Microbiol 14:815–820

Turra A, Petracco M, Amaral, ACZ, Denadai MR (2014) Population biology and secondary production of the harvested clam *Tivela mactroides* (Born, 1778) (Bivalvia, Veneridae) in Southeastern Brazil. Mar Ecol, 1–14, DOI: 10.1111/maec.12137

UNCSD United Nations Conference on Sustainable Development (2012) The future we want. A/RES/66/288, United Nations General Assembly, Rio de Janeiro, RJ, Brazil. p 53

Underwood AJ (1997) Experiments in Ecology: Their Logical Design and Interpretation Using Analysis of Variance. Cambridge University Press, Cambridge

Vieira RHSF, Atayde MA, Carvalho EMR, Carvalho FCT, Fonteles Filho AA (2008) Contaminação fecal da ostra *Crassostrea rhizophorae* e da água de cultivo do estuário do Rio Pacoti (Eusébio, Estado do Ceará): Isolamento e identificação de *Escherichia coli* e sua susceptibilidade a diferentes antimicrobianos. Braz J Vet Res Anim Sci 45(3):180–189

Villeneuve JP, Carvalho FP, Fowler SW, Cattini C (1999) Levels and trends of PCBs, chlorinated pesticides and petroleum hydrocarbons in mussel from the NW Mediterranean coast: comparison of concentrations in 1973/1974 and 1988/1989. Sci Total Environ 237(238):57–65

Wenzl T, Simon R, Alklam E, Kleiner J (2006) Analytical methods for polycyclic aromatic hydrocarbons (PAHs) in food and the environment needed for new food legislation in the European Union. TrAC Trends Anal Chem 25(7):716–725

Wingqvist GO, Drakenberg O, Slunge D, Sjöstedt M, Ekbom A (2012) The role of governance for improved environmental outcomes: Perspectives for developing countries and countries in transition. Swedish Environmental Protection Agency, Report 6514, Stockholm, Sweden. p 60

Woodhead RJ, Law RJ, Matthiessen P (1999) Polycyclic aromatic hydrocarbons in surface sediments around England and Wales, and their possible biological significance. Mar Pollut Bull 38(9):773–790

Yunker MB, Macdonald RW, Vingarzan R, Mitchell RH, Goyette D, Sylvestre S (2002) PAHs in the Fraser river basin: a critical appraisal of PAH ratios as indicators of PAH source and composition. Org Geochem 33(4):489–515

10

Assessment of microbiological quality of some drinks sold in the streets of Dhaka University Campus in Bangladesh

Mahbub Murshed Khan[1], Md Tazul Islam[2], Mohammed Mehadi Hassan Chowdhury[3*] and Sharmin Rumi Alim[1]

Abstract

Background: Various kinds of fruit juices and sherbets sold by the street vendors are widely consumed in Dhaka city. These street vended fruit juices and sherbets are usually high in microbial loads. So, the present study was undertaken to assess the microbiological load, possible risk factors and identity of freshly squeezed juices and sherbets and their safety for human consumption in terms of pathogens.

Result: For the study purpose papaya juice, sugarcane juice, tukmaria sherbet, lemon sherbet and wood apple sherbet were taken as samples. The study showed a high microbial load in the drinks. The range of average total viable count (microbial load) and total coliforms were 7.7×10^3 - 9×10^8 cfu/ml and 210–1100 cfu/100 ml, indicated the heavy presence of microorganisms in all the drinks analyzed in this study. The study revealed that tukmaria sherbet was most contaminated with a count of 9×10^8 cfu/ml. The least contamination was observed in lemon sherbet. A count of 1.98×10^6 cfu/ml was observed in papaya juice and 3.4×10^5 cfu/ml was in wood apple sherbet. Total coliforms were present in all samples and average count for total coliforms was high in tukmaria sherbets than others. Various pathogenic species of bacteria such as *Proteus* sp., *Enterobacter* sp, *E. coli*, *Shigella* sp, *Citrobacter* sp, *Vibrio* sp, *Yersinia* sp and *Hafnia* sp were isolated from the juices and sherbets. Unhygienic water for dilution, dressing with ice, prolonged use without refrigeration, insanitary surroundings, raw materials, chemical properties, equipment, fruit flies and airborne dust are the risk factors of contamination.

Conclusion: It was revealed that consumption of these street drinks were harmful for people. These drinks can cause various food borne illnesses. People should avoid these drinks.

Keywords: Dhaka city; Fruit juice; Sherbet; Pathogen; Food borne illness

Background

In developing countries, various kinds of drinks sold by the street vendors are widely consumed by millions of people. These drinks are available in metropolitan and other cities of Bangladesh. In Bangladesh, the drinks usually consumed by the people at the streets or from roadside shops are various carbonated soft drinks, tea, coffee, fruit juices and sherbets. But the major drinks sold by the street vendors and found to consume by the people at the street are various kinds of fruit juices and sherbets. According to FAO, street food is a food obtained from a street side vendor, often from a make-shift or portable stall (FAO 2007).

During summer season (during the months of March through August) a huge section about of the population of all income and age groups consumes these fresh pressed and squeezed juices and sherbets (Ahmed et al. 2009). These drinks are sold all the year round except winter season. Unpasteurized juices are preferred by the consumers because of the "fresh flavor" attributes and low cost. They are simply prepared by mechanically squeezing fresh fruits or may be extracted by water. The final product is an unfermented, clouded, untreated drink, ready for consumption. The drinks are nutritious for man. These are also good medium for the growth of many microorganisms some of which may be pathogenic to man. The pathogenic organism can cause various

* Correspondence: md.mehadihassanchy@yahoo.com
[3]Department of Microbiology, Noakhali Science and Technology University, Noakhali, Bangladesh
Full list of author information is available at the end of the article

food borne diseases. Consumption of fresh fruit juices are increasing day by day. In addition to their increasing popularity in consumption patterns, fresh fruits and vegetables have also become increasingly important vehicles in food borne disease statistics (Sivapalasingam et al. 2004).

Consumption of fresh fruits and their juice provides potential health benefits to the general population (Alothman et al. 2009). Juices are fat-free, nutrient dense beverages that are rich in vitamins, minerals and naturally occurring phytonutrients that contribute to good health. It is good for health only when it is free from pathogenic microorganisms.

Contamination of food products can result in many health problems ranging from mild bloating and gas to serious incidents of food poisoning and dehydration. Unsafe and non hygienic fruits consumptions cause serious outbreaks of food borne illness (Sivapalasingam et al. 2004). There have been some notable outbreaks of illness in recent years that demonstrate the increasingly important role of fresh fruits and vegetables in food borne disease (Parish 1997; Sandeep et. al. 2001). There are several reports of illnesses due to the food borne diseases associated with the consumption of fruit juices at several places around the globe (Mosupye and Holy 2000; Muinde and Kuria 2005; Chumber et al. 2007). Usually raw materials, equipments, hand of the handlers, containers etc. are responsible for contamination. Contamination from raw materials and equipments, additional processing conditions, improper handling, prevalence of unhygienic conditions contributes substantially to the entry of bacterial pathogens in juices prepared from these fruits or vegetables (Oliveira et al. 2006; Nicolas et al. 2007). Water used for juice preparation can be a major source of microbial contaminants such as total coliforms, fecal coliforms, fecal streptococci etc. (Tasnim et al. 2010).

Street vended fruit juices and sherbets are found to sell in almost all areas of Dhaka city. There is a high demand of fresh fruit juices and sherbets in Dhaka city. Most of the time consumers are not conscious about the safety, quality, and hygiene of the drinks. It can be potential factor for food borne diseases. In view of the threat posed by the bacterial pathogens in street vended drinks and the flourishing demands of these drinks, the present study was undertaken to assess the bacteriological load and identity of freshly squeezed juices and sherbets and their safety for human consumption in terms of bacterial pathogens.

Methods
Sample collection
Two different types of drinks were taken as sample for the study purpose; these were 1) Fruit juice and 2) Sherbet. There were two varieties of fruit juices namely papaya juice and sugarcane juice, and three types of sherbets namely tukmaria sherbet, lemon sherbet and wood apple sherbet. 3 samples of each type, total 15 samples were collected for test. The drinks were collected in sterile containers from the different vendors of five different places of Dhaka University Campus. The vendors were the drink and juice producers and they produced these in their makeshift stalls with unhygienic raw materials such as water, ice, syrups etc. and environment. The containers were kept in ice-boxes and transported to the laboratory.

Sample processing
The samples were in liquid form, so further processing was not required. One ml of liquid sample was taken and transferred into sterilized cotton plugged test tubes containing 9 ml of 0.1% peptone water. Then they were mixed thoroughly by shaking 20 times. This time the solution was allowed to stand for 5-10minutes.Thus initial dilution of homogenate was prepared and from this homogenization further serial dilution were prepared.

Bacteriological studies
Pour plate method was used for the determination of total viable count using total plate count agar for water. This media is unique and nonselective to count bacteria and fungi. The samples were diluted at 10 fold dilution up to 10^{-6} according American public Health Association (APHA) sample dilution guidelines (APHA 1992). Two plates from every dilution for each temperature at (36 ± 2)°C for (44 ± 4) hrs for bacteria and (22 ± 2)°C for (68 ± 4) hrs for yeast & molds were incubated. To identify the isolated bacteria; cultural, morphological and biochemical characteristics were studied carefully by using "Bergey's Manual of Determinative Bacteriology", 9th Ed. (Holt et al. 1994). Different types of selective agar were used for isolation and identification of bacterial colonies using streak plate method. The instructions for using of media were followed by guidelines of producing company. MacConkey agar was used to isolate Gram-negative enteric bacteria. To isolate and differentiate between *Shigella* sp. and *Salmonella* sp. SS agar was used because it was a differential media. TCBS agar was used as a selective media to isolate *Vibrio* spp. EMB (Eosin methylene blue) agar was used for the isolation of *E.coli* as a selective medium for *E.coli* and *Proteus*. The bacterial colonies grown on different types of medium were collected and maintained in Nutrient slant agar for further analysis. Some biochemical tests such as Kliger's Iron Agar (KIA) test, Motility test, Imvic test, Indole test, Urea (MIU), test Catalase and Oxidase tests were performed for the identification of bacterial isolates. MPN method is used for the quantitative estimation for

coliforms using three test tubes where inoculums size were 10, 1, 0.1 ml. MacConkey broth and BGLB (Brilliant green lactose broth) were used in this experiment (APHA 1992).

Results and discussion

Colony morphology, phenotypic and biochemical traits of the isolates

After 24 hours of incubation, MPN test determined the estimation of total coliforms using MPN Chart. Yellow and green colonies on TCBS agar were primarily considered as *Vibrio* sp. Colorless, transparent, with a black center colonies in SS Agar were primarily considered as *Salmonella* sp. Green metallic sheen indicated the presence of *E. coli* on EMB plate. Biochemical tests gave the confirmation for the identification of bacterial isolates.

Microbiological study

From the study it was clear that all the juices and sherbets contain a significant amount of microorganisms. The average total viable count (microbial load) showed the presence of microbes in all the drinks analyzed in this study in the range of 7.7×10^3 - 9×10^8 cfu/ml. The study showed that tukmaria sherbet was most contaminated with a count of 9×10^8 cfu/ml. Microbial load of sugarcane juice was 2.55×10^6 cfu/ml. A count of 1.98×10^6 cfu/ml was observed in papaya juice and 3.4×10^5 cfu/ml was in wood apple sherbet. The least contamination was observed in lemon sherbet. The presence of total coilifroms in all drinks implied a negative relation with food quality and safety. The average total viable count and total coliforms count from the drinks were presented in the Table 1 and the isolates identified in the samples were listed in the Table 2.

Ahmed et al. (2009) found a range of 3×10^2 to 9.6×10^8 microorganisms in freshly squeezed fruit juices sold in Dhaka City as well as Rashed et al. (2013) found a total microbial load within a range of 1.9×10^3 to 2.8×10^7 cfu/ml and in the present study, microbial load recorded within the range of 10^3 to 10^8, which is quite similar to the findings of Ahmed et al. (2009) and Rashed et al. (2013) (Ahmed et al. 2009; Rashed et al. 2013).

Table 1 Average estimation of Total viable count and total coliforms from the drinks were given bellow

Type of drinks	Number of samples	Average total viable count (CFU/ml)	Average MPN (CFU/100 ml)
Sugarcane juice	3	2.55×10^6	290
Tukmaria sherbet	3	9×10^8	1100
Lemon sherbet	3	7.7×10^3	210
Wood apple sherbet	3	3.4×10^5	240
Papaya juice	3	1.98×10^6	460

Table 2 Identified bacterial isolates in sample drinks

Sugarcane juice	Tukmaria sherbet	Lemon sherbet	Wood apple sherbet	Papaya juice
Proteus	*Proteus*	*Proteus*	*Proteus*	*Proteus*
Enterobacter	*Enterobacter*	*Yersinia*	*Hafnia*	*Enterobacter*
E. coli	*E. coli*		*E. coli*	*E. coli*
Shigella	*Shigella*			*Shigella*
Citrobacter	*Salmonella*			*Vibrio*
Vibrio	*Vibrio*			
	Aeromonas			

The highest contamination was recorded in tukmaria sherbet. It might be due to the raw materials used in this drink. Bacterial load in lemon sherbet was comparatively low. It may be due to low pH of lemon sherbet. The low pH of fruit juices greatly limits the number and the type of bacteria that can survive or grow at that pH (Prescott et al. 2002).

From the data presented in the current study, it could be concluded that the microbiological quality of most of the drinks collected from different areas of Dhaka University Campus were not satisfactory as fecal coliforms like *E. coli*, *Enterobacter* and other pathogens like *Shigella*, *Vibrio* were detected from the samples and all the samples contained higher load of microbes than the Gulf standard (Gulf Standards 2000) showed in table 3. The lack of knowledge on safe fruit juice preparation as well as the contamination source can contribute to the pathogens in prepared juices.

The presence of these microbes in food can be linked to a number of risk factors such as improper handling and processing, use of contaminated water during washing and dilution, cross contamination from rotten fruits and vegetables, or the use of dirty processing utensils like knife, flies and trays which was so relevant with the study of Khalil et al. (Khalil et al. 1994). This might also implicate the processing and rinsing water as possible sources of contamination of street foods sold by street vendors (Nwachukwu et al. 2008). The ice and water added during preparation were likely to provide possible sources of additional bacterial contamination (Bryan et al. 1998, 1992). Furthermore, these fruit juices were

Table 3 The recommended Microbiological standards for any fruit juices sold in the Gulf region (Gulf standards, 2000)

Test	Total aerobic bacterial count (cfu/mL)	Total coliforms count (cfu/mL)	Yeasts and molds (cfu/mL)
Maximum count anticipated	5.0×10^3	10	100
Maximumcount permitted	5.0×10^4	100	1.0×10^3

left in ambient temperature which may have led to the proliferation of contaminating bacteria resulting in increased bacterial counts (Bryan et al. 1977, 1992).

In the study various pathogenic bacteria were isolated from the juices and sherbets which may cause a serious outbreak in Bangladesh as like as USA where 21 fruit juice-associated outbreaks of illness were reported to the Centers for Disease Control and Prevention (CDC) from 1995 through 2005 (Vojdani et al. 2008). The presence of *Salmonella* was recorded in tukmaria sherbet. It was possible that *Salmonella* may have gained entry through water because vendors do not use boiled water and WASA water in Dhaka city is highly contaminated (Parveen et al. 2008) and this water is commonly used for diluting juices and washing utensils used for washing and preparing juices, alternately, the possibility of contamination of fruits through improperly treated irrigation water cannot be ruled out; survival and entry of enteropathogens including *Salmonella* have been shown in crops, irrigated with contaminated sewage (Ryu and Beuchat 1998). This organism causes typhoid and paratyphoid in human.

Present studies clearly indicated the presence of different types of fecal coliforms namely, *Citrobacter, Enterobacter* and *Escherichia coli* in all the drinks except lemon sherbet. The presence of coliforms in fruit juice is not allowed by safe food consumption standard (Andres et al. 2004). These organisms are highly pathogenic and may cause serious diseases in human beings. *Citrobacter* causes urinary tract infections, infections in gall bladder, middle ear etc. *Enterobacter* are responsible for urinary tract infections and hospital sepsis etc. *Escherichia coli* cause diarrhea, urinary infections, pyogenic infections and septicemia etc. (Ananthanarayan and Jayaram 1996; Falagas et al. 2007; Samonis et al. 2009). *Shigella* causes dysentery that result in the destruction of the epithelial cells of the intestinal mucosa in the cecum and rectum. Some strains produce the enterotoxin shiga toxin which causes hemolytic uremic syndrome (Koster et al. 1978). Several species of *Vibrio* are pathogenic (C. Michael 2010) and a significant number of fruit and vegetable borne outbreaks have been reported (Faruque et al. 1998). Most disease-causing strains are associated with gastroenteritis, but can also infect open wounds and cause septicemia. *Vibrio cholerae* infects the intestine and increases mucous production causing diarrhea and vomiting which result in extreme dehydration and, if not treated, death (Howard-Jones 1984).

But to prevent such contamination in fruit juice drinks a Good Hygienic Practice (GHP) as defined in the Codex document on "General Principles of Food Hygiene" in combination with HACCP is the basis for safe food production (Codex Alimentarius, 1997). Government Health Agencies must adopt measures to educate the vendors about food safety and hygienic practices and to develop a regular monitoring system. Robust traceability would greatly improve the epidemiological investigation, identification of causal or contributory factors and the control of suspected incidents of foodborne illness. There is a need for improved surveillance systems on food-borne pathogens, on food products and on outbreaks so that comparable data are available from a wider range of countries.

Conclusions

Low price and availability make the street vended juices and sherbets highly popular to general people. But these drinks will be beneficial if these are free from contamination. The current study revealed that all the studied fruit juices and sherbets had a higher microbial load than the specification set for fruit juices in some parts around the world. Regular monitoring of the quality of fruit juices for human consumption is recommended to avoid any future bacterial pathogen outbreak.

Abbreviations
C.F.U.: Colony forming unit; MPN: Most probable number; FAO: Food and agricultural organization.

Competing interest
The authors declare that they have no competing interest.

Authors' contributions
All the authors contributed equally to the preparation of this manuscript. All authors read and approved the final manuscript.

Author details
[1]Institute of Nutrition and Food Science, University of Dhaka, Dhaka, Bangladesh. [2]Department of Food Technology and Nutrition Science, Noakhali Science and Technology University, Noakhali, Bangladesh. [3]Department of Microbiology, Noakhali Science and Technology University, Noakhali, Bangladesh.

References
Ahmed MSU, Nasreen T, Feroza B, Parveen S (2009) Microbiological Quality of Local Market Vended Freshly Squeezed Fruit Juices in Dhaka City, Bangladesh. Bangladesh J Sci Ind Res 44(4):421–424

Alothman M, Bhat R, Karim AA (2009) Effects of radiation processing on phytochemicals and antioxidants in plant produce. Trends Food Sci Tech 20:201–212

American Public Health Association (APHA) (1992) Standard Methods for the Examination of Water and Wastewater, 18th edn. APHA, Washington, D.C.

Ananthanarayan R, Jayaram Paniker CK (1996) Text Book of Microbiology, 5th edn. Orient Longman Limited, Chennai, pp 40–43, 250-261

Andres SC, Giannuzzi L, Zaritzky NE (2004) The effect of temperature on microbial growth in apple cubes packed in film and preserved by use of orange juice. Int J Food Sci Technology 39(9):927–933

Bryan FL (1977) Diseases transmitted by foods contaminated with wastewaters. J. Food Protect 40:45–56

Bryan FL, Teufel P, Riaz S, Roohi S, Qadar F, Malit Z (1992) Hazards and critical control points of vending operations at a railway station & bus station in Palaitar. J Food Prot 55:534–541

Bryan FL, Michanie SC, Alvarez P, Paniagua A (1998) Critical control points of street vended foods in the Dominican Republic. J Food Prot 51:373–383

Chumber SK, Kaushik K, Savy S (2007) Bacteriological analysis of street foods in Pune. Indian J Public Health 51(2):114–116

Codex Alimentarius (1997) Recomended International Code of Practice, General Principles of Food Hygiene. Supplement to volume 1 B. Joint FAO/WHO Food Standards Programme, FAO, Rome

Falagas ME, Rafailidis PI, Kofteridis D, Virtzili S, Chelvatzoglou FC, Papaioannou V, Maraki S, Samonis G, Michalopoulos A (2007) Risk factors of carbapenem-resistant Klebsiella pneumoniae infections: a matched case–control study. J Antimicro Chemother 60:1124–1130

FAO (2007). The informal food sector. http://www.informalfood.unibo.it 2007-11-23.

Faruque SM, Albert MJ, Mekalanos JJ (1998) Epidemiology, genetics and ecology of oxigenic Vibrio cholerae. Microbiol Mol Biol Rev 62:1301–1314

Gulf Standards (2000) Microbiological Criteria for Foodstuffs-Part 1. GCC, Riyadh, Saudi Arabia

Holt JG, Krieg NR, Sneath PHA, Staley JT, Williams ST (1994) Bergey's manual of determinative bacteriology, 9th edn. Lippincott Williams & Wilkins, Baltimore, MD

Howard-Jones N (1984) Robert Koch and the cholera vibrio: a centenary. BMJ 288 (6414):379–381. doi:10.1136/bmj.288.6414.379

Khalil K, Lindblom GB, Mazhar K, Kaijser B (1994) Flies and water as reservoirs for bacterial enteropathogens in urban and rural areas in and around Lahore Pakistan. Epidemiol Infect 113:435–444

Koster F, Levin J, Walker L, Tung KSK, Gilman RH, Rahaman MM, Majid A, Islam S, Williams RC (1978) Hemolytic uremic syndrome after shigellosis: relation to endotoxemia and circulating immune complexes. N Engl J Med 298:927–933

Hogan CM (2010) Bacteria, Encyclopedia of Earth, Sidney Draggan and C.J. Cleveland, National Council for Science and the Environment, NCSE, Washington DC

Mosupye FM, Holy AV (2000) Microbiological hazard identification and exposure assessment of street food vending in Johannesburg, South Africa. Int J Food Microbiol 61:137–145

Muinde OK, Kuria E (2005) Hygienic and sanitary practices of vendors of street foods in Nairobi, Kenya. Afr J Food Agri Nutri Dev 5(1):1–13

Nicolas B, Razack BA, Yollande I, Aly S, Tidiane OCA, Philippe NA, Comlan DS, Sababénédjo TA (2007) Street Vended foods improvement: contamination mechanism and application of food safety objective strategy: Critical Review. Pak J Nutr 6(1):1–10

Nwachukwu E, Ezeama CF, Ezeanya BN (2008) Microbiology of polyethylene-packaged sliced watermelon (Citrullus lanatus) sold by street vendors in Nigeria. Afr J Microbiol Res 2:192–195

Oliveira ACG, Seixas ASS, Sousa CP, Souza CWO (2006) Microbiological evaluation of sugarcane juice sold at street stands and juice handling conditions in São Carlos, São Paulo, Brazil. Cad Saude Publica 22(5):1111–1114

Parish ME (1997) Public health and non pasteurized fruit juices. Crit Rev Microbiol 23:109–119

Parveen S, Ahmed MSU, Nasreen T (2008) Microbial Contamination of water in around Dhaka city. BJSIR 43(2):273–276

Prescott LM, Harly JP, Kleen DA (2002) Food Microbiolgy, 5th edn. McGraw Hill Book Co., New York, pp 352–627

Rashed N, Md A, Md U, Azizul H, Saurab KM, Mrityunjoy AM, Majibur R (2013) Microbiological study of vendor and packed fruit juices locally available in Dhaka city, Bangladesh. Int Food Res J 20(2):1011–1015

Ryu JH, Beuchat LR (1998) Influence of acid tolerance responces on Survival, growth and crossprotection of Escherichia Coli 0157: Media and fruit juices. Int. J Food Microbiol 45:185–193

Samonis G, Karageorgopoulos DE, Kofteridis DP, Mattaiou DK, Sidiropoulou V, Maraki S, Falagas ME (2009) Citrobacter infections in a general hospital: characteristics and outcomes. Eur J Clin Microbiol Infect Dis 28(1):61–68

Sandeep MD, Waker A, Abhijit G (2001) Microbiological Analysis of street vended fresh squeezed carrot and kinnow-mandarin juices in Patiala city, India. Internet J Food Safety 3:1–3

Sivapalasingam S, Friedman CR, Cohen L, Tauxe RV (2004) Fresh produce: a growing cause of outbreaks of food borne illness in the United States, 1973 through 1997. J Food Prot 67:2342–2353

Tasnim F, Hossain MA, Nusrath S, Hossain MK, Lopa D, Haque KMF (2010) Quality Assessment of Industrially Processed Fruit Juices Available in Dhaka City, Bangladesh. Malaysia J Nutri 16(3):431–438

Vojdani JD, Beuchat LR, Tauxe RV (2008) Juice-associated outbreaks of human illness in the United States, 1995 through 2005. J Food Prot 71(2):356–364

11

The GM Contamination Register: a review of recorded contamination incidents associated with genetically modified organisms (GMOs), 1997–2013

Becky Price[1] and Janet Cotter[2]*

Abstract

Background: Since large-scale commercial planting of genetically modified (GM) crops began in 1996, a concern has been that non-GM crops may become contaminated by GM crops and that wild or weedy relatives of GM crops growing outside of cultivated areas could become contaminated. The GM Contamination Register contains records of GM contamination incidents since 1997 and forms a unique database. By the end of 2013, 396 incidents across 63 countries had been recorded.

Results: Analysis of the Register database reveals rice has the highest number of GM contamination incidents of all crops (accounting for a third of incidents), despite there being no commercial growing of GM rice anywhere in the world. The majority of these incidents derive from two distinct cases of contamination of unauthorised GM rice lines, LLRICE from the USA and BT63 rice from China. Maize accounts for 25% of GM contamination incidents, whilst soya and oilseed rape account for approximately 10% of incidents. Although factors such as acreage grown, plant biology, designation as a food or non food crop and degree of international trading can potentially affect the frequency and extent of contamination, it is not possible to determine which are dominant.

The Register records a total of nine cases of contamination from unauthorised GM lines, i.e. those at the research and development stage with no authorisation for commercial cultivation anywhere in the world. An important conclusion of this work is that GM contamination can occur independently of commercialisation. Some of these cases, notably papaya in Thailand, maize in Mexico and grass in USA have continued over a number of years and are ongoing, whilst other contamination cases such as Bt10 maize and pharmaceutical-producing GM crops occur only with a single year. The route(s) of contamination are often unclear.

Conclusions: The detection of GMO contamination is dependent on both routine and targeted monitoring regimes, which appears to be inconsistent from country to country, even within the EU. The lack of an analytical methodology for the detection of GM crops at the field trial stage (i.e. pre-commercialisation) can hamper efforts to detect any contamination arising from such GM lines.

Keywords: Genetically modified; Contamination; Gene flow; Crops; Food; Feed; Field trial

* Correspondence: janet.cotter@greenpeace.org
[2]Greenpeace Resaerch Laboratories, Innovation Centre Phase 2, University of Exeter, Exeter EX4 4RN, UK
Full list of author information is available at the end of the article

Background

Large scale commercial planting of genetically modified (GM) crops began in 1996. Alongside concerns regarding the long term health, environmental and socio-economic impacts of these crops, a specific concern has been that of contamination of non-GM crops by GM crops and also of relatives established outside planted areas. Concerns regarding GM contamination of non-GM crops include loss of markets, particularly those requiring "GM-free" products; future supply of non-GM seed (especially for seed saved from open pollinated crops) and possible introgression (spreading) of the GM trait into both wild and feral populations of crop relatives (Marvier and van Acker 2005; Mellon and Rissler 2004; Bauer-Panskus et al. 2013).

Contamination of non-GM crops by GM crops has occurred, and there are several well-documented cases, e.g. Starlink corn (Fox 2001), Liberty Link rice (LLRICE; US Food and Drug Administration 2006). Reviews of GM contamination by Marvier and van Acker (2005) and Bauer-Panskus et al. (2013) considered that there are many factors contributing to the likelihood of GM contamination, and that containment of any contamination can be problematic, even unlikely. For example, Marvier and van Acker (2005) considered of the movement of transgenes a virtual certainty via routes such as gene flow during cultivation, reintroduction of transgenes from volunteer and feral crop populations, seed transport (including between continents) and human error such as co-mingling of seed. Bauer-Panskus et al. (2013) considered that domestication and ease of hybridization with wild or weedy relatives important factors. Possible environmental consequences of introgression of transgenes into wild or feral population include increased invasiveness as a result of a fitness enhancing trait such as insect resistance (Snow et al. 2003; Samuels 2013), reduced capacity for co-existence of GM and non-GM crops (Marvier and van Acker 2005) and hampering future breeding efforts (Bauer-Panskus et al. 2013). However, Ellstrand (2012) found that transgene escape to wild varieties was rare, but that transgenes tended rather to move into other varieties of the same species.

The United Nations (UN) Cartagena Protocol on Biosafety to the Convention on Biological Diversity partly addresses the control of living genetically modified organisms (GMOs) (termed living modified organisms within the convention), covering seeds (UN Cartagena Protocol on Biosafety 2014). In 2002, this established an advance informed agreement procedure meaning that only those living GMOs (primarily seeds) with formal approval from the receiving country's national body can be imported. In addition to the UN Cartagena Protocol, many governments around the world, including the European Union (EU), have established national regulatory regimes that issue formal authorisation (or, in the USA, deregulation) for any deliberate release of GMOs into the environment (i.e. outside of laboratory containment) and marketing as seed, human food and animal feed products. Specific GMOS may be authorised either as experimental field trials (e.g. for testing of agronomic traits), for commercial cultivation and for marketing as human food and/or animal feed. Authorisations can be granted for a GMO to be marketed without cultivation, e.g. if the GMO is to be imported into a country, but not grown. Databases of authorisations for the cultivation and marketing of GMOs are available both globally (Center for Environmental Risk Assessment 2014) and also for certain regions such as the EU (EC 2014a). In contrast, databases for experimental field trials of GMOs are held by national or regional authorities, e.g. for the USA (Information Systems for Biotechnology 2014), Australia (Office of the Gene Technology Regulator 2014), the European Union (EC 2014b). In many countries, such databases do not exist, or are not publically available.

The presence of globally unauthorised GMOs (i.e. those without any authorisation for cultivation or marketing anywhere in the world) and unapproved GMO crops (i.e. those without authorisation for cultivation and/or marketing in that country or region) in food/feed or seed can be grounds for either product recalls or rejection of imports at national borders. Where labelling of GMO ingredients is mandatory (e.g. the EU for food/feed), the presence of GMO ingredients in unlabelled foodstuffs above a prescribed threshold (0.9% for the EU, EC 2003) is also grounds for either rejection of imports or product withdrawal. Labelling is not required for GM non food products (e.g. biofuels or materials such as cotton) in the EU (EC 2003). Global organic farming standards require that both seed and food are free from GMOs (International Foundation for Organic Agriculture Movements 2002) and many national organic farming bodies have defined thresholds of the maximum permitted adventitious GMO content (usually between 0.1% and 0.9%). The presence of GMOs above permitted thresholds can result in loss of organic certification.

Methodologies for the detection of GMOs fall into two broad categories: protein-based methods which target the novel protein(s) produced by the GMO and deoxyribonucleic acid (DNA)-based methods which target the inserted genetic construct(s). Protein-based methods typically employ an enzyme-linked immunosorbent assay (ELISA). Commercially-available ELISA kits comprise either qualitative immunochromotographic (lateral flow) strips, which can be used both in the field and laboratory, or well plates enabling quantitative analysis. In contrast, DNA-based methods utilise polymerase chain reaction (PCR) and require a specialised laboratory set up. The EC Joint Research Centre (JRC) has developed and validated methods for the detection and quantification of individual GMO events (EC JRC 2010, 2014a) and run international collaborative programmes, predominantly with governmental laboratories, to increase harmonisation and standardisation of methodologies for analysing GMOs (EC JRC 2014b). In practice, protein-based methods (especially qualitative assays) tend to be used to screen for the presence of GMOs (Ermolli et al. 2006), whilst DNA-based methods offer more robust analysis (Miraglia et al. 2004).

Despite the concerns regarding GM contamination, there is no systematic global monitoring or recording of GM contamination incidents. In the EU region, however, recording of GM contamination incidents in food/feed products is afforded by the European Commission (EC) Rapid Alert System on Food and Feed (RASFF) (EC RASFF 2014). RASFF members consist of the European Union countries together with Iceland, Liechtenstein, Norway and Switzerland. To date, the RASFF provides the most comprehensive data set of GM contamination in food/feed provided by a national or regional regulatory body. Along with other food contaminants, the RASFF records positive results on its website (EC RASFF 2014) for unapproved GM ingredients in food/feed products both from within the EU region and imports into the EU region. Outwith the RASFF, individual incidents of GM contamination are generally reported via national governments, non governmental organisations (NGOs) and/or the media.

In the absence of global systematic monitoring of GM contamination incidents, GeneWatch UK and Greenpeace International established the GM Contamination Register (hereafter termed "The Register") to record these incidents. The website (GM Contamination Register 2014) is a searchable database used by individuals, public interest groups and governments. It is recognised by the UN Biosafety Clearing-House, an information exchange facility for the Cartagena Protocol. The Register now holds a database of recorded incidents of GM contamination over 17 years, from 1997 to the present time. This paper presents an analysis of the recorded incidents from 1997 to the end of 2013. We investigate the frequency of GM contamination incidents over this period, the principal crops and countries associated with these incidents and focus on contamination arising from unauthorised GM crops. Finally, we investigate the contribution of the contribution of analytical methodologies and monitoring regimes to the detection of GM contamination incidents.

Methodology

The GM Contamination Register is based on individual incidents reported by governmental and inter-governmental authorities, non governmental organisations (NGO)s and/or the media. Monitoring English speaking media, science journals and internet-based news groups provides alerts to these incidents. Organisations testing for the presence of GMOs (such as NGOs) are invited to send their results to the Register. Coverage is global, and GM animals and plants are included. GM microbes are not excluded, although there have been no recorded incidences of contamination by these organisms to date. The Register commenced in 2005, was backdated to 1997. The Register is still active although only those incidents recorded to the end of 2013 are included here.

Each report of a contamination incident is evaluated for its dependability. The Register includes only incidents where the contamination can be verified to a reasonable degree of certainty. That is, either an announcement by a producer of a GM crop, a governmental body or from an NGO based on the identification of the GM crop resulting from PCR analysis. The PCR methodologies need to be accredited to the International Organization for Standardization (ISO), i.e. ISO standard 17025 (ISO 2005) and this usually entails contracting analysis to a commercial or governmental laboratory. GM contamination incidents based on protein testing are excluded from the Register. Incidents reported via the RASFF are regarded as verified as they originate from government bodies. Similarly, those published in peer reviewed journals are also regarded as verified. Where the incident is announced by an NGO, copies of PCR results are requested as verification.

Where an incident cannot be verified from original or governmental sources, it is accepted for the Register if the reportage is via a reputable newspaper or news agency, preferably an international agency such as Reuters or Agence France-Presse. Such cases are reviewed after one year and internet searches are made for further evidence. In cases where the contamination findings are refuted or reasonable doubt cast on them, the incident is removed from the Register until any new evidence emerges.

For inclusion in the GM Contamination Register, incidents must meet at least one of the following criteria:

- a breach of national or international law; sale of food/feed derived from GMOs that does not comply with regional or national labelling regulations,
- presence of unauthorised or unapproved GMO traits in crops and/or seed batches or, for approved GMO traits, presence above applicable thresholds,
- illegal plantings of GMOs or unauthorised releases to the environment or food/feed chain.
- establishment of feral population(s) of a GM crop or presence of the genetic insert within wild or feral populations, including wild or weedy relatives.

If a GM incident has been verified and fulfils one of the criteria, brief details of the contamination incident (generally what, where, when and, if known, the source and route of contamination) are entered into the database. Only incidents that have been publicly documented are recorded on the Register. The Register does not, and cannot, record incidents that remain undetected. Therefore, those reported here represent the minimum number of GM contamination incidents.

The nature of GM contamination is such that once a single contamination incident has been identified, testing and monitoring can quickly result in the identification of several more contamination incidents involving the same GM line, often within the same month or year. In addition, international trading of agricultural commodities can result in repeated findings of contamination by the same GM line in many countries within the same year. These are termed 'repeated contamination' incidents and could skew the Register towards particular GM lines that are frequently tested for, or countries that conduct more thorough monitoring. To avoid this, the Register only records a new incident each time a GM line is found to be present within a country in a particular year. If the same GM line repeatedly contaminates within the same country and in the same year, no new incident is recorded and details are added to the existing incident instead. This enables the Register to indicate the extent of spread, both geographically and over time, of any given contamination episode whilst minimising any skew towards repeated contamination incidents.

Results and Discussion

As of 31 December 2013, the Register had recorded 396 incidents since 1997 (i.e. over a period of 17 years) and across 63 different countries (Additional file 1). Since 2000, there have been more than 10 incidents per year and, since 2005, more than 20 incidents per year (Figure 1). For 2006, there is a sharp spike in the number of incidents to nearly

60. This spike primarily relates to the discovery and spread of GM rice from the USA, which accounted for 28 incidents that year.

GM contamination incidents by country

The distribution of the total number of GM contamination incidents over the 17 year period by country displays an exponential decay type of pattern with a sharp initial decrease from the highest number of incidents per country and a long tail of countries with a much lower number of incidents (Figure 2). 50% of the total number of incidents are accounted for by 11 countries, each with more than 12 incidents. This group is dominated by RASFF countries, together with North America and Australia. In contrast, 28 countries (just under half the total number of countries) with 3 or less reported incidents account for just under 10% (9.1%). The remainder of the RASFF countries, several Asian countries, South Korea, China and Mexico recorded between 3 and 12 incidents per country.

The 4 countries reporting the highest number of incidents were Germany, USA, France and United Kingdom, three of which contribute to the RASFF. Together, these countries make up 27% of the total number contamination incidents. This does not correspond to the countries with the highest acreage of GM crops, which are USA, Brazil, Argentina and Canada (International Service for the Acquisition of Agri-biotech Applications 2013).

The RASFF database accounts for a significant number of incidents in the Register, approximately 60% (244 out of 396) (Figure 3). The importance of recording repeated contamination events associated with the same GM line, within the same year and same country as one incident is illustrated by RASFF. If each notification for RASFF were included on the Register, they would account for nearly 80% of the total number of incidents (585 out of 736), dominating the Register and producing a skew

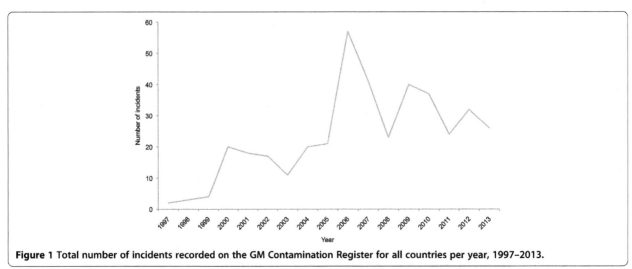

Figure 1 Total number of incidents recorded on the GM Contamination Register for all countries per year, 1997–2013.

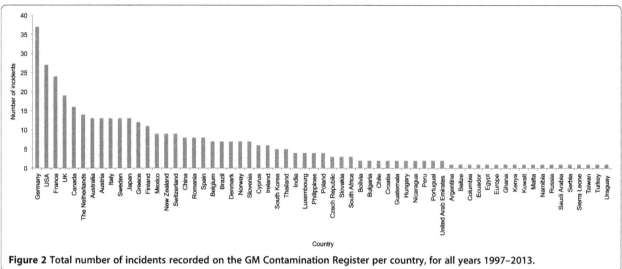

Figure 2 Total number of incidents recorded on the GM Contamination Register per country, for all years 1997–2013.

towards these repeated contamination events. The prominence of GM contamination incidents recorded for the EU via RASFF does not necessarily reflect a higher degree of GM contamination. It is more likely that there is routine monitoring and a mechanism for reporting such incidents.

GM contamination by crop

Rice is associated with the highest number of GM contamination incidents of all crops (Table 1), accounting for about a third of the total number of incidents. This is despite a global absence of any commercial cultivation of GM rice. There is a sharp peak in the number of incidents of GM rice contamination for 2006/7 (Figure 4).

The majority of these incidents derive from two distinct cases of contamination from unauthorised GMOs, one in the USA (LLRICE) and one in China (Bt63) and are further discussed in the section on globally unauthorised GM crops below. GM maize gives rise to the second most number of contamination incidents, accounting for a quarter of incidents (Table 1) with consistent contamination of between 5 and 10 incidents per year since 1999 (Figure 4). Soya and oilseed rape (canola) are the next most frequently associated with GM contamination, each accounting for approximately 10% of incidents (Table 1) with between 0 and 10 incidents per year (Figure 4).

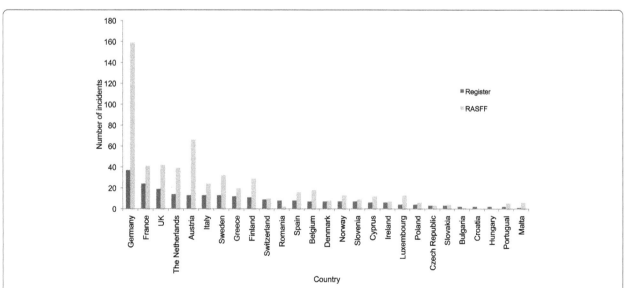

Figure 3 Comparison of GM Contamination Register with the EC RASFF database by country, for all years 1997–2013. EC RASFF database records notifications for each incident, including those relating to a single GM line in the same country and year. In the GM Contamination Register, these are recorded as a single incident, resulting is a lower total no. of incidents for each RASFF country. This prevents a skew of the data towards RASFF countries and the reportage of repeated incidents.

Table 1 Total number of incidents recorded on the GM Contamination Register for each crop for all countries and all years, 1997–2013

Crop	No. of incidents	% Total no. of incidents
Rice	134	34
Maize	98	25
Oilseed rape/canola	40	10
Soybean	37	9
Flax	26	6.5
Papaya	18	4.5
Cotton	14	3.5
Fish	5	1.3
Grass	4	1
Pigs	4	1
Sugar beet	4	1
Arabidopsis thaliana	3	0.75
Potato	2	0.5
Alfalfa	1	0.25
Plum	1	0.25
Tomato	1	0.25
Wheat	1	0.25
Zucchini	1	0.25
Pollen in honey*	1	0.25
Cherry, kiwi & olive trees**	1	0.25
Total	396	

*Pollen from GM corn, oil seed rape & soya was found in honey imported into Switzerland (Greenpeace 2009).
**Experimental field trials of these GM trees exceeded the length of permit in Italy and are recorded together as one incident (Genetic Rights Foundation 2012).

The acreage of a GM crop has potential to influence the number of contamination incidents because it could increase the opportunities for cross-pollination and co-mingling with non-GM seed/grain. International trading of a crop could also affect the number of incidents as many reported GM contamination incidents originate from national border checks. Maize, soya and oilseed rape are all major commodity crops and also the principal GM feed/food crops that are commercially grown. GM varieties are widely grown in many countries where they are authorised for cultivation. For example, in the USA, where GM crops were first cultivated, GM crop acreage accounted for 93% of all soya and 90% of corn in 2013 (USDA Economic Research Service 2014). A pilot study in the US found that DNA derived from GM crops was found in 50% of non-GM varieties of corn and soy, and 83% of non-GM oilseed rape (Mellon and Rissler 2004). Although these findings are not regarded as contamination because no regulations have been infringed, it demonstrates that GMO traits are widespread in the USA. These GMO traits can enter into non-GM crops destined for export to countries/regions where they may not be approved or require labelling (e.g. the EU). GM cotton was also widely grown in the USA in 2013, accounting for 90% of the crop acreage but only accounting for < 4% of contamination incidents. This may be because it is a non food crop, and therefore not subject to labelling in the EU or inclusion in the RASFF. In contrast, the acreage of GM rice grown is extremely small, amounting only to field trials. However, rice is a highly traded commodity crop, with approximately 1.2 million tonnes imported into the EU during 2012/3 (EC 2013). In addition, there has been specific monitoring for unauthorised GM varieties in many countries post 2006, when the unauthorised cases of contamination were discovered. These two factors could have contributed to the high number of contamination incidents associated with GM rice, despite the global absence of commercial growing.

Plant biology, and specifically the tendency to outcross, is a factor that could affect the number of GM contamination incidents. GM maize is reported as "medium to high" risk and oilseed rape as "high" risk for pollen mediated gene flow (Eastham and Sweet 2002) and this may also contribute to the high number of

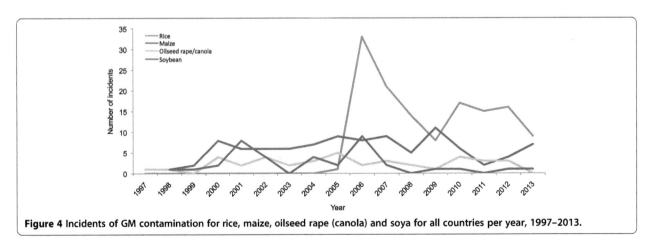

Figure 4 Incidents of GM contamination for rice, maize, oilseed rape (canola) and soya for all countries per year, 1997–2013.

incidents reported for these two crops, but wouldn't be a strong factor for cotton and soya. Gene flow from GM rice, despite being regarded as a largely self pollinating crop, has been recorded, albeit at low levels and over short distances (Rong et al. 2007). Although factors such as acreage grown, plant biology, designation as a food or non food crop and degree of international trading can affect the number of contamination incidents recorded, it is not possible to determine from these data which are dominant. Indeed, for rice, the specific monitoring for unauthorised varieties of GM rice may to be an additional factor.

For maize and canola, there has been much international interest in specific cases of GM contamination, notably those associated with Starlink maize, maize in Mexico and establishment of feral GM oilseed rape populations. GM Starlink maize was only authorised for use in animal feed in 1998 (Bucchini and Goldman 2002, US Environmental Protection Agency 2008) but was found in USA food products intended for human consumption in 2000 by a coalition of non governmental groups (Fox 2001). The contamination persisted in the US at least until 2003, despite efforts to recover Starlink seed (Marvier and van Acker 2005). Starlink maize was the first contamination case with international ramifications. Contamination incidents were reported in 6 further countries (Bolivia, Canada, Egypt, Guatemala, Japan and South Korea) during both 2000 and 2001. Starlink was also reported in food aid to Bolivia in 2004 (Meridian Institute 2002) and present in foods in Saudi Arabia in 2009 and 2010 (Elsanhoty et al. 2013).

In Mexico, GM contamination of open-pollinated traditional varieties or landraces of maize was initially reported in a highly controversial publication in 2001 (Quist and Chapela 2001). The contamination was thought to originate from imports of GM maize from North America. The GM contamination was not detected in a 2003/4 study (Ortiz-Garcia et al. 2005) but later studies (Piñeyro-Nelson et al. 2009; Dyer et al. 2009) did find GM contamination of landraces. It is now generally accepted that GM transgenes are present in at least some maize landraces in Mexico (Dalton 2008, Snow 2009).

The establishment of feral populations of GM oilseed rape in both Canada and Japan has recently been reviewed in-depth by Bauer-Panskus et al. (2013). They describe that, in Canada, two types of feral GM oilseed rape populations conferring tolerance to two different herbicides, glufosinate ammonium (trade name: Liberty or Basta) and glyphosate (trade name: Roundup) occur in provinces where they are widely grown, or through which the grain is transported. In places, these have hybridized to be tolerant to both herbicides. In Japan, cultivation of GM oilseed rape is minor, but much oilseed rape is imported from Canada. Feral populations of GM oilseed rape tolerant to glufosinate, glyphosate or both have been reported in and around Japanese ports.

Globally Unauthorised GMOs

Several of the GM contamination incidents recorded on the Register are associated with "unauthorised" GM lines. That is, lines that were not authorised for commercial cultivation or marketing anywhere in the world at the time of the contamination incident but were experimental, or in development. Thus, risk assessments for food/feed and environmental safety have generally not been performed. Contamination incidents involving unauthorised GM lines tend to have a higher media profile than incidents originating from approved GM lines because of their disruptive effects on trade, as well as increased food and environment safety concerns (Holst-Jensen 2008) because of the lack of risk assessment. Hence, these incidents are further described.

GM Rice

Two unauthorised GM contamination cases account for 128 of the 133 rice incidents on the Register. These are "LLRICE" from the United States and "Bt63 rice" from China.

In April 2005, Greenpeace announced that they had evidence of illegal sales and cultivation of unauthorised GM rice in the Hubei province of China (Greenpeace 2005; Zi 2005). Field trials of insect resistant GM rice Bt63 had been carried out in this province for a number of years (Tu et al. 2000, Huang et al. 2005). In 2006, GM rice continued to contaminate food products in China, including baby foods produced by Heinz (Greenpeace 2006a). Bt63 also started to be detected in Europe, with Greenpeace, Friends of the Earth and European governments announcing it had been identified in imported products in Austria, France, UK and Germany. This led to 5 recorded incidents on the Register for 2006 (Figure 5). In 2007, the contamination was also identified in products imported into further countries with reports coming from Japan, Cyprus, Germany, Greece, Italy, Sweden and the UK.

In 2008, the EU Commission announced emergency controls on all rice products from China to prevent imports of unauthorised GM rice (EC 2008, Huggett 2008) with further measures imposed 3 years later (EC 2011). These restrictions require consignments to be certified as not containing GM rice and imports to be subjected to sampling and document checks at the EU port of entry. Although Figure 5 does not show any reduction in GM contamination incidents as a result of these EU measures, it is possible that this is an artefact of increased monitoring, which could have resulted in a greater detection of GM contamination.

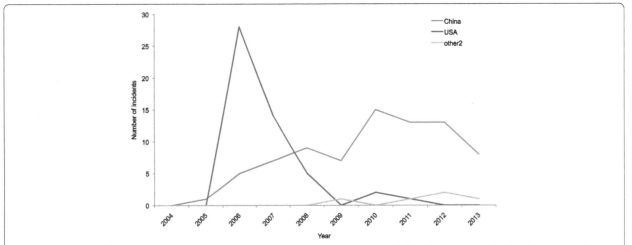

Figure 5 Frequency of GM contamination of rice 2005–2013. Number of incidents recorded on the GM Contamination Register for all countries per year for three unauthorised GM rice cases: US LLRICE; Chinese Bt63 and "other", comprising principally GM contaminated basmati rice either in, or imported from, India and Pakistan.

In the USA, two varieties of glufosinate-tolerant GM rice, LLRICE62 and LLRICE06, were authorised for cultivation and marketing (i.e. afforded deregulated status) in 2000 (Center for Environmental Risk Assessment 2014). These GM varieties have never been grown commercially. A separate glufosinate-tolerant line of GM rice, LL601, was under development by Bayer Crop Science until development (and field trials) ceased in 2001. In August 2006, the 2005 crop of (non-GM) rice in the USA was contaminated with LLRICE601 (US Food and Drug Administration 2006). Following the announcement, LLRICE601 was given deregulated status in the US (Center for Environmental Risk Assessment 2014). However, at the time of contamination it was unauthorised for cultivation and marketing anywhere the in the world. Further, in March 2007, the United States Department of Agriculture (USDA) confirmed that rice line had become contaminated with yet another unauthorised GM rice line LLRICE604, also developed by Bayer Crop Science (USDA 2007). However, LL604 does not appear to have caused as widespread contamination as LL601.

The Register data show how, following the initial announcement in the USA, LLRICE (predominantly LL601) was discovered in 28 countries around the world including EU member states, Sierra Leone, Ghana, Philippines, Kuwait, Mexico and United Arab Emirates, all during 2006. The following 3 years show a rapid fall off rate (Figure 5) but despite this, 6 years elapsed from discovery of the contamination to the last recorded case in 2011.

The two cases of GM rice contamination both caused disruption to international trade but have different histories of contamination rates (Figure 5). Chinese Bt63 continues to be detected in EU imports from its initial discovery in 2005 through to 2013. US LLRICE, however, saw a rapid decline in incidents from 2006–2009

with few cases after this date. This may possibly represent an ease of halting the GM contamination and/or differing monitoring regimes, although these data do not allow further speculation.

There are 5 incidents of GM contamination of rice other than LLRICE and Bt63 (Figure 5). Of these, 4 incidents relate to GM contaminated basmati rice either in, or imported from, India and Pakistan and 1 relates to GM contaminated animal feed imported from the USA.

GM grass
Between 1999 and 2005, Scotts company conducted field trials of GM creeping bentgrass (*Agrostis stolonifera*), tolerant to the herbicide glyphosate in the USA. However, the USA Animal and Plant Health Inspection Service (APHIS) found that the company failed to prevent escape of the GM grass during 2003 (US APHIS 2004) and had failed to remove immature seed heads in 2005 (US APHIS 2007a). Gene flow from GM bentgrass can occur over large distances and has been detected over 21 km from source (Watrud et al. 2004). The GM bentgrass was found to have established in non agronomic habitats up to 3.8 km away from an Oregon field test site (Reichman et al. 2006). GM bentgrass remains present in uncultivated habitats in Oregon (Charles 2011) and has hybridised with a naturalised grass species (Zapiola and Mallory-Smith 2012). This is an ongoing case of GM contamination and, because of the nature of grass pollen and seeds, it may prove difficult to contain and eradicate (Reichman et al. 2006, Charles 2011).

Pharmaceutical-producing GM crops
In 2001, ProdiGene conducted field trials in Iowa, USA of GM maize producing a vaccine against a pig disease. In September 2002, government inspectors discovered

volunteer maize plants growing in a soybean field that was used as a test site for the experimental GM maize in 2001 (US APHIS 2002). Additionally, because the GM maize volunteers may have pollinated neighbouring commercial maize fields, all maize seed and plant material within 1,320 ft (400 m) of the previous year's test plot was destroyed (US APHIS 2002). At a second site in the same year, APHIS destroyed 500,000 bushels (estimated value US$2.7 million) after soya from a field containing volunteer GM maize was harvested and mixed with other soya (US APHIS 2002).

In 2004, volunteer GM pharmaceutical corn was again found growing and flowering within the fallow zone surrounding a field trial site and in a nearby sorghum field. The company destroyed all maize volunteers within the 1 mile (1,610 m) isolation zone, and to plough under the sorghum field (US APHIS 2007b).

A further incident of contamination involving pharmaceutical-producing GM crops occurred in March 2008. The Belgium competent authorities notified the EC RASFF that traces of a protein derived from unauthorised GM *Arabidopsis thaliana* in a food supplement. The transgenic protein was intended to be used in the measurement of vitamin B12 in laboratory medicine (Bor et al. 2003), and not intended as a food supplement. The product had been imported into Belgium from Denmark.

GM Papaya

Several different lines of GM papaya resistant to papaya ringspot virus (PRSV) have been under development in tropical regions. However, only two have been authorised for commercial growing, both in the USA (Center for Environmental Risk Assessment 2014). A total of 18 cases of GM contamination have, however, been recorded on the Register. These relate to contamination arising from the development of GM papaya resistant to PRSV in Taiwan and Thailand. Papaya is a widely grown in these two countries, both domestically and for export. Neither country has ever authorised GM papaya for cultivation.

In 2003, GM papaya resistant to PRSV was discovered growing on Taiwanese farms and for sale in local market places (Yu-Tzu 2003). Japan halted Taiwanese papaya imports and the Taiwanese government imposed a ban on GM papaya. The government warned that anyone found growing or selling such fruits could face a fine, and reiterated this warning in 2006 (Taiwan News 2006). The origin of the experimental GM papaya is thought to be government agricultural laboratories developing GM papaya resistant to PRSV.

A similar case of GM papaya contamination arose in Thailand. GM papaya has not been authorised for cultivation in Thailand. In 2004, however, Thai government officials confirmed a Greenpeace finding of the presence of GM papaya in famers' fields (Davidson 2008). A government research station had been developing GM papaya resistant to PRSV, including conducting field trials, but was also selling (non-GM) seed to farmers. Government testing found a sample of sold seed which contained GM papaya. This selling of GM contaminated seed is thought to be the source of the contamination (Davidson 2008). 85 north-eastern farmers were found to have inadvertently grown GM papaya in 2004 (Greenpeace 2006b) when the government ordered the destruction of all GM papaya plants, including the field trial. In 2005, GM papaya was found in more regions of Thailand (Greenpeace 2006b).

GM papaya was also detected in Thai exports of papaya. The Register records 11 RASFF alerts for GM papaya entering the EU since 2007, with 4 incidents recorded in 2012, and 12 in 2013. GM contamination in Thai papaya appears to be either ongoing. With the exception of Japan, no countries outwith RASFF have reported GM contamination from Thai papaya imports but it is not clear what, if any, inspection regimes exist.

GM Maize Bt10

The first case of an unauthorised GM line contaminating international food supplies was recorded in 2005. Syngenta had produced two types GM maize lines, Bt10 and Bt11 with two traits, insect-resistance and herbicide tolerance. Only one – Bt11 maize – was eventually commercialised. However, Syngenta's quality control systems had apparently failed to detect Bt10 seeds had been contaminating commercial stocks of Bt11 for 4 years prior to the discovery (Macilwain 2005). Despite this, the case gave rise to only 4 incidents on the Register, primarily because the EU commission made a report on behalf of the whole of the EU. No incidents are recorded for Bt10 contamination beyond 2005.

GM Linseed (Flax)

GM linseed (or flax) line FP967, named "Triffid", was developed by a public research institution in Canada. It was authorised for commercial use in both Canada and the United States in the late 1990's. However, Triffid GM seed was never sold for commercial production and, in 2001, Triffid was de-registered and it was believed that all known stocks had been identified and destroyed (Canadian Grain Commission 2010, Ellstrand 2012).

In September 2009, Germany found unauthorised genetic material in linseed imported from Canada (EC RASFF 2009), which was subsequently confirmed as Triffid. This led to further testing for GM contamination of linseed and in 2009 the Register recorded a total of 16 incidents in Canada and 15 in the RASFF countries. In 2010, there were 8 incidents recorded on the RASFF and a further one recorded in 2011. Positive test results were found in Canada 2009–2011. The levels of GM contamination were reported as falling during these years (Franz-Warkentin 2011) and ongoing efforts include encouraging producers to grow only licenced flax varieties that are free of GM contamination (Flax Council of Canada 2014).

GM wheat

One case of GM wheat contamination has been reported (Table 1). In 2013, an Oregon farmer noticed that some volunteer wheat did not die-back as expected when sprayed with glyphosate. Analysis by Oregon State University and the USDA determined these volunteers to be a glyphosate tolerant GM wheat line, MON71800 (USDA 2013). Although MON71800 was passed for food and feed consumption in the USA in 2004 (Center for Environmental Risk Assessment 2014), it is included here as the development of the GM wheat line was discontinued in 2004 (Monsanto 2004) and the application for commercialisation (deregulation) was withdrawn. Exports of wheat from the USA where disrupted whilst testing took place (USDA 2013), but no detection of GM wheat has been reported by importer countries. To date, the route of this contamination is unknown but it appears to be an isolated incident.

GM animals

The Register records 4 incidents (one per year 2001–2005) where experimental GM pigs have entered into the food or feed supply unauthorised. These are either accidentally comingled with non-GM livestock at the abattoir, sometimes due to mislabelling or, in one case, deliberately stolen (Westphal 2001). Rather than an environmental release, these are an unauthorised entry of a GMO into food or feed. No incidents of contamination of GM fish or insects have been recorded on the Register to date. On the whole, terrestrial GM livestock are considered less prone to escape and establishment of feral populations than GM fish or insects (National Research Council 2004). Especially, experimental GM livestock tend to be kept in animal houses, rarely allowed into an open environment and each individual is tagged and monitored. However, these cases suggest that these too can cause GM contamination.

Detection and monitoring of GM contamination

The Register records only reported and/or verified incidents of GM contamination. As such, the Register does not represent all contamination incidents that have taken place globally. There are undoubtedly other GM contamination incidents that remain undetected or unreported. This could be because it is not possible to detect the GMO, that monitoring for GM contamination either does not take place or is insufficient to detect all or many GM contamination incidents and/or that GM contamination is not reported.

The RASFF is the only database which systematically records GM contamination events but is only for the EU region. The high number of incidents documented on the Register from RASFF (approximately 60% of the total) demonstrates its effectiveness. Nevertheless, the RASFF database is the product of national monitoring, and dependent on the effectiveness of national monitoring regimes for monitoring GM contamination. The high number of events recorded by the RASFF countries of Germany, France and United Kingdom may reflect more efficient monitoring programmes, rather than a high number of products or imports containing unapproved GMOs.

GMO detection relies on the establishment of an analytical methodology specific to that GMO, e.g. the PCR methodologies collated by the EC JRC (EC JRC 2010, 2014a). Any contamination by unauthorised GMOs, such as those in experimental field trials, may not be readily detectable, as most (if not all) national regulatory authorities do not require companies to submit an analytical methodology for detection of the test GM crop when seeking approval for field trials. The lack of an analytical methodology can make detection of GM contamination extremely difficult, if not impossible. PCR screening for elements that are common to many GM crops (e.g. the cauliflower mosaic virus promoter, CaMV 35S) can indicate of whether GM material is present. However, it does not allow identification of the gene that confers the trait, which is necessary to identify the specific GM line. For example, whilst detection methodologies are now available for both LLRICE and Chinese Bt63 (EC JRC 2014a), these were not available until after the GM contamination had been reported. For the cases of pharmaceutical-producing GM crops, no detection methodology is available. Thus, it would not be possible for any third party to monitor for the presence/absence of these GM crops in subsequent crop harvests, if contamination was suspected.

The Register reports cases of contamination from unauthorised GM crops. These cases include LLRICE, Bt63 rice, grass, pharmaceutical maize, papaya, and Bt10 maize all of which were at the research or development stage, and Triffid linseed and wheat which had undergone some safety assessment, but were not authorised for cultivation. These nine cases clearly show that GM contamination can occur independently of commercialisation, e.g. escapes from field trials or illegal plantings.

Once a GM contamination incident has been detected, this often leads to monitoring specifically for the associated GM line and can result in a high number of incidents immediately after the initial discovery (e.g. LLRICE). This can lead to a sharp increase in the number of contamination incidents per year for that crop, e.g. as seen for rice in 2006/7. The pattern of GM contamination incidents varies a great deal between cases. For example, the cases of Bt10 and all pharmaceutical GM crops cases only extend over a year. Starlink, LLRICE and FP967 flax show a sharp decrease after 2/3 years but with ongoing isolated incidents of GM contamination. It is not clear why some cases of contamination fall rapidly and others not so rapidly. Governments and trading bodies wish reduce any economic impact from GM contamination as soon as possible. In some cases at least, this has evidently spurred governments and companies to make efforts to eliminate contamination, such as the destruction of GM papaya in Thailand, or licensing GM-free flax varieties. Internationally-traded commodity crops, therefore, are likely to receive the most efforts to reduce GM contamination.

The seed supply system is a factor in determining the persistence of GM contamination. GM contamination of maize landraces in Mexico may be difficult to eradicate because these varieties are dominantly open-pollinated, where seed is saved from harvest for next year's planting. The GM traits could even introgress through these populations (Dyer et al. 2009). In contrast, hybrid seed systems are more centrally controlled. Hybrid seed is purchased each year and hybrid seeds are produced via crossing of specified varieties. This allows the testing of the parental varieties for the absence GM material prior to breeding, each year if necessary.

GM grass in the western USA could represent the first case where a GM crop has entered into wild or naturalized populations, and it remains to be seen whether it will introgress through those populations.

Routes of GM contamination

Many cases of GM contamination are not investigated, but those from unauthorised events have been investigated the most. Cross pollination is not the sole route of GM contamination. A number of routes have been identified including escape of seeds (grass), illegal plantings (Bt63 rice), incorrect labelling (pigs) and volunteers from previous year's crops have all been identified as the cause of GM contamination. It's also possible that a GM contamination case arises from two or more routes. Often, the route of GM contamination is not clear. For example, Bayer was unable to identify the route of contamination responsible for the LLRICE case, calling it "an act of God" (Weiss 2006).

Finally, it is beyond the scope of the Register or this paper to speculate about health or environmental implications from these specific contamination cases. However, if GMOs that have not undergone any food/feed or environmental risk assessments enter the food/feed chain or environment, this gives regulators cause for concern.

Conclusions

Nearly 400 incidents of GM contamination have been recorded on the Register since 1997. There does not appear to be an overall pattern relating to any one particular factor. Instead, factors could include, but may not be limited to global acreage, international trade, plant biology and monitoring frequency. Experimental GM livestock have, on occasion, entered either human food or animal feed.

All three principal, commercially grown GM food and feed crops (oilseed rape, soya and maize) have been associated with GM contamination incidents over the past 17 years. There have also been nine cases of contamination associated with GM lines with no authorisation for cultivation anywhere in the world, mostly at the research and development stage. An important conclusion of this work is that GM contamination can occur independently of commercialisation. Indeed, GM lines of rice, the crop associated with the highest number of incidents, has never been grown commercially. The detection of GMO contamination is dependent on both routine and targeted monitoring regimes, which appears to be inconsistent from country to country, even within the EU. The lack of an analytical methodology for the detection of GM crops at the field trial stage (i.e. pre-commercialisation) can hamper efforts to detect any contamination arising from such GM lines.

Additional file

Additional file 1: GM contamination database - analysis.xlsb.

Abbreviations

APHIS: Animal and Plant Health Inspection Service; DNA: Deoxyribonucleic acid; EC: European commission; ELISA: Enzyme-Linked Immunosorbent Assay; EU: European Union; EPA: Environmental Protection Agency; GM: Genetically modified; GMO: Genetically modified organism; ISO: International organisation for standardization; JRC: Joint research centre; LL: Liberty link; NGO: Non-governmental organisation; PRSV: Papaya ringspot virus; PCR: Polymerase chain reaction; RASFF: Rapid alert system for food and feed; UN: United Nations; USA: United States of America; USDA: United States Department of Agriculture.

Competing interests

Neither author has competing financial interests although both authors are employed by NGOs which oppose the environmental release of GM crops.

Authors' contributions

The GM Contamination Register is maintained by B.Price and part funded by Greenpeace. Both authors contributed equally to the preparation of this manuscript. Both authors read and approved the final manuscript.

Author details

[1]Consultant for GeneWatch UK, 60 Lightwood Road, Buxton SK17 7BB, UK. [2]Greenpeace Resaerch Laboratories, Innovation Centre Phase 2, University of Exeter, Exeter EX4 4RN, UK.

References

Bauer-Panskus A, Breckling B, Hamberger S, Then C (2013) Cultivation-independent establishment of genetically engineered plants in natural populations: current evidence and implications for EU regulation. Environ Sci Eur 25:34. doi:10.1186/2190-4715-25-34

Bor MV, Fedosov SN, Laursen NB, Nexø E (2003) Recombinant human intrinsic factor expressed in plants is suitable for use in measurement of vitamin B12. Clin Chem 49:2081–2083

Bucchini L, Goldman LR (2002) Starlink corn: a risk analysis. Environ Health Perspect 110:5–13

Canadian Grain Commission (2010) Background information on genetically modified material found in Canadian flaxseed. www.grainscanada.gc.ca/gmflax-lingm/pfsb-plcc-eng.htm. Accessed 14 May 2014

Center for Environmental Risk Assessment (2014) GM Crop Database. Center for Environmental Risk Assessment (CERA). ILSI Research Foundation, Washington D.C. http://cera-gmc.org/GMCropDatabase. Accessed 14 May 2014

Charles D (2011) Scientist in the middle of the GM-Organic Wars. Science 332:168

Dalton R (2008) Modified genes spread to local maize. Nature 456:149

Davidson SN (2008) Forbidden fruit: transgenic papaya in Thailand. Plant Physiol 147:487–493

Dyer GA, Serratos-Hernández JA, Perales HR, Gepts P, Piñeyro-Nelson A, Chávez A, Salinas-Arreortua N, Yúnez-Naude A, Taylor JE, Alvarez-Buylla ER (2009) Dispersal of transgenes through maize seed systems in Mexico. PLoS One 4:e5734

Eastham K, Sweet J (2002) Genetically Modified Organisms (GMOs): the Significance of Gene Flow Through Pollen Transfer. Expert's Corner Series. European Environment Agency, Copenhagen. http://www.eea.europa.eu/publications/environmental_issue_report_2002_28 Accessed 14 May 2104

EC (2003) Regulation (EC) No 1829/2003 of the European Parliament and of the Council of 22 September 2003 on genetically modified food and feed. O J Eur Union L 268:1–23

EC (2008) Commission requires certification for Chinese rice products to stop unauthorised GMO from entering the EU. http://europa.eu/rapid/pressReleasesAction.do?reference=IP/08/219&type=HTML&aged=0&language=EN&guiLanguage=en. Accessed 14 May 2014

EC (2011) EC O J Eur Union L 343:140–147

EC (2013) EU rice trade 2012/13. http://ec.europa.eu/agriculture/cereals/trade/rice/2012-13_en.pdf. Accessed 14 May 2014

EC (2014a) EU Register of authorised GMOs. http://ec.europa.eu/food/dyna/gm_register/index_en.cfm. Accessed 18 Sept 2014

EC (2014b) Deliberate release into the environment of GMOs for any other purposes than placing on the market (experimental releases). http://gmoinfo.jrc.ec.europa.eu/. Accessed 18 Sept 2014

EC JRC (2010) Compendium of reference methods for GMO analysis. http://ihcp.jrc.ec.europa.eu/our_activities/gmo/gmo_analysis/compendium-reference-methods-gmo-analysis. Accessed 18 Sept 2014

EC JRC (2014a) GMO Methods: EU Database of Reference Methods for GMO Analysis. http://gmo-crl.jrc.ec.europa.eu/gmomethods/. Accessed 14 May 2014

EC JRC (2014b) GMO Capacity Building. http://ihcp.jrc.ec.europa.eu/our_activities/gmo/gmo_capacity_building. Accessed 18 Sept 2014

EC RASFF (2009) Notification detail - 2009.1171. https://webgate.ec.europa.eu/rasff-window/portal/index.cfm?event=notificationDetail&NOTIF_REFERENCE=2009.1171 Accessed 14 May 2014

EC RASFF (2014) Rapid Alert System for Food and Feed (RASFF). http://ec.europa.eu/food/food/rapidalert/index_en.htm. Accessed 14 May 2014

Ellstrand NC (2012) Over a Decade of Crop Transgenes out-of-Place. In: Wozniak CA, McHughen A (ed) Regulation of Agricultural Biotechnology: The United States and Canada. Springer, London

Elsanhoty RM, Al-Turki AI, Ramadan MF (2013) Prevalence of genetically modified rice, maize, and soy in Saudi food products. Appl Biochem Biotechnol 171:883–899

Ermolli M, Fantozzi A, Marini M, Scotti D, Balla B, Hoffmann S, Querci M, Paoletti C, van den Eede G (2006) Food safety: screening tests used to detect and quantify GMO proteins. Accred Qual Assur 11:55–57

Flax Council of Canada (2014) GM Flax Update January 18 2014. http://flaxcouncil.ca/files/web/NEWS%20RELEASE%20-%20Flax%20Council%20of%20Canada%20Urges%20Producers%20to%20Declare%20Unlicensed%20Flax%20Varieties%20-%2001.28.14.pdf. Accessed 14 May 2014

Fox JL (2001) EPA re-evaluates StarLink license. Nat Biotechnol 19:11

Franz-Warkentin P (2011) Flax Industry Sees "Good Progress" Against Triffid. Manitobar Co-operator, Canada. http://www.manitobacooperator.ca/daily/flax-industry-sees-good-progress-against-triffid-2. Accessed 14 May 2014

Genetic Rights Foundation (2012) Illegal field trials of GM cherry, kiwi and olive trees exposed in Italy by the Genetic Rights Foundation. http://www.fondazionedirittigenetici.org/fondazione/en/displaynews_en.php?id=19. Accessed 14 May 2014

GM Contamination Register (2014). http://www.gmcontaminationregister.org Accessed 14 May 2014

Greenpeace (2005) Greenpeace discovers illegal GE rice in China. http://www.greenpeace.org/international/en/news/features/scandal-greenpeace-exposes-il/. Accessed 14 May 2014

Greenpeace (2006a) Illegal genetically engineered rice found in Heinz baby food in China. http://www.greenpeace.org/eastasia/press/releases/20060314-heinz-rice-cereal/ Accessed 14

Greenpeace (2006b) Contamination by genetically engineered papaya in Thailand. http://www.greenpeace.org/international/Global/international/planet-2/report/2006/6/GEpapayaThailand.pdf. Accessed 14 May 2014

Greenpeace (2009) Des pollens transgéniques dans le miel d'importation. http://www.greenpeace.org/switzerland/fr/publications/actualites/agriculture/pollens-transgeniques-miel/ Accessed 14

Holst-Jensen A (2008) GMO testing - trade, labeling or safety first? Nat Biotechnol 26:858–859

Huang J, Hu R, Rozelle S, Pray C (2005) Insect resistant GM rice in farmers' fields: assessing productivity and health effects in China. Science 308:688–690

Huggett B (2008) EU to monitor for Chinese GM rice. Nat Biotechnol 26:478

Information Systems for Biotechnology (2014) USDA Field Tests of GM Crops. http://www.isb.vt.edu/search-release-data.aspx. Accessed 17 Sept 2014

International Foundation for Organic Agriculture Movements (2002) Position on genetic engineering and genetically modified organisms. http://www.ifoam.org/en/position-genetic-engineering-and-genetically-modified-organisms. Accessed 14 May 2014

International Service for the Acquisition of Agri-biotech Applications (2013) Brief 46: Executive Summary. Global Status of Commercialized Biotech/GM Crops: 2013. http://www.isaaa.org/resources/publications/briefs/46/executivesummary/default.asp Accessed 14 May 2014

ISO (2005) ISO/IEC 17025:2005 General requirements for the competence of testing and calibration laboratories. http://www.iso.org/iso/catalogue_detail.htm?csnumber=39883. Accessed 18 Sept 2014

Macilwain C (2005) US launches probe into sales of unapproved transgenic corn. Nature 434:423

Marvier M, van Acker RC (2005) Can crop transgenes be kept on a leash? Front Ecol Environ 3:92–100

Mellon M, Rissler J (2004) Gone to Seed: Transgenic Contaminants in the Traditional Seed Supply. Union of Concerned Scientists. http://www.ucsusa.org/food_and_agriculture/our-failing-food-system/genetic-engineering/gone-to-seed.html. Accessed 18 Sept 2014

Meridian Institute (2002) Illegal Genetically Engineered Starlink Corn Contaminates Food Aid. http://www.merid.org/Content/News_Services/Food_Security_and_AgBiotech_News/Articles/2002/06/11/Illegal_Genetically_Engineered_Starlink_Corn_Contaminates_Food_Aid.aspx. Accessed 16 Oct 2014

Miraglia M, Berdal KG, Brera C, Corbisier P, Holst-Jensen A, Kok EJ, Marvin HJP, Schimmel H, Rentsch J, van Rie JPPF, Zagon J (2004) Detection and traceability of genetically modified organisms in the food production chain. Food Chem Toxicol 42:1157–1180

Monsanto (2004) Monsanto to realign research portfolio, development of roundup ready wheat deferred. http://news.monsanto.com/press-release/monsanto-realign-research-portfolio-development-roundup-ready-wheat-deferred. Accessed 18 Sept 2014

National Research Council (2004) Bioconfinement of animals: fish, shellfish, and insects. In: Biological confinement of genetically engineered organisms. US National Academies Press, Washington DC, pp 130–158

Office of the Gene Technology Regulator (2014) Genetically modified organisms - field trial sites. Department of Health, Australian Government. http://www.ogtr.gov.au/internet/ogtr/publishing.nsf/content/map. Accessed 18 Sept 2014

Ortiz-Garcia S, Ezcurra E, Schoel B, Acevedo F, Soberon J, Snow AA (2005) Absence of detectable transgenes in local landraces of maize in Oaxaca, Mexico (2003–2004). Proc Natl Acad Sci U S A 102:12338–12343

Piñeyro-Nelson A, Van Heerwaarden J, Perales HR, Serratos-Hernández JA, Rangel A, Hufford MB, Gepts P, Garay-Arroyo A, Rivera-Bustamante R, Alvarez-Buylla ER (2009) Transgenes in Mexican maize: molecular evidence and methodological considerations for GMO detection in landrace populations. Mol Ecol 18:750–761

Quist D, Chapela I (2001) Transgenic DNA introgressed into traditional maize landraces in Oaxaca, Mexico. Nature 414:541–543

Reichman JR, Watrud LS, Lee EH, Burdick CA, Bollman MA, Storm MA, King GA, Mallory-Smith C (2006) Establishment of transgenic herbicide-resistant creeping bentgrass (Agrostis stolonifera L.) in nonagronomic habitats. Mol Ecol 15:4243–4255

Rong J, Lu B-R, Song Z, Su J, Snow AA, Zhang X, Sun S, Chen R, Wang F (2007) Dramatic reduction of crop-to-crop gene flow within a short distance from transgenic rice fields. New Phytol 173:346–353

Samuels J (2013) Transgene flow from Bt brinjal: a real risk? Trends Biotechnol 31:333–334

Snow A (2009) Unwanted transgenes re-discovered in Oaxacan maize. Mol Ecol 18:569–571

Snow AA, Pilson D, Riesberg LH, Paulsen MJ, Pleskac N, Reagon MR, Wolf DE, Selbo SM (2003) A Bt transgene reduces herbivory and enhances fecundity in wild sunflower. Ecol Appl 13:279–86

Taiwan News (2006) Genetically modified papaya said unsafe. http://www.taiwannews.com.tw/etn/news_content.php?id=105252. Accessed 14 May 2014

Tu J, Zhang G, Datta K, Xu C, He Y, Zhang O, Khush G, Datta SK (2000) Field performance of transgenic elite commercial hybrid rice expressing Bacillus thuringiensis delta-endotoxin. Nat Biotechnol 18:1101–1104

UN Cartagena Protocol on Biosafety (2014). http://bch.cbd.int/protocol/ Accessed 14 May 2014

US APHIS (2002) Noncompliance history: ProdiGene. http://www.aphis.usda.gov/wps/portal/aphis/resources/enforcement-actions?1dmy&urile=wcm%3apath%3a%2Faphis_content_library%2Fsa_our_focus%2Fsa_biotechnology%2Fsa_compliance_and_inspections%2Fct_compliance_history. Accessed 14 May 2014

US APHIS (2004) Noncompliance history: The Scotts Company. http://www.aphis.usda.gov/wps/portal/aphis/resources/enforcement-actions?1dmy&urile=wcm%3apath%3a%2Faphis_content_library%2Fsa_our_focus%2Fsa_biotechnology%2Fsa_compliance_and_inspections%2Fct_compliance_history. Accessed 14 May 2014

US APHIS (2007a) Noncompliance history: The Scotts Company LLC. http://www.aphis.usda.gov/wps/portal/aphis/resources/enforcement-actions?1dmy&urile=wcm%3apath%3a%2Faphis_content_library%2Fsa_our_focus%2Fsa_biotechnology%2Fsa_compliance_and_inspections%2Fct_compliance_history. Accessed 14 May 2014

US APHIS (2007b) Noncompliance history: ProdiGene. http://www.aphis.usda.gov/wps/portal/aphis/resources/enforcement-actions?1dmy&urile=wcm%3apath%3a%2Faphis_content_library%2Fsa_our_focus%2Fsa_biotechnology%2Fsa_compliance_and_inspections%2Fct_compliance_history. Accessed 14 May 2014

US Environmental Protection Agency (2008) Starlink™ Corn Regulatory Information. http://www.epa.gov/pesticides/biopesticides/pips/starlink_corn.htm. Accessed 14 May 2014

US Food and Drug Administration (2006) Statement on report of bioengineered rice in the food supply. http://www.fda.gov/Food/FoodScienceResearch/Biotechnology/Announcements/ucm109411.htm. Accessed 14 May 2014

USDA (2007) Report of LibertyLink Rice Incidents. http://www.aphis.usda.gov/newsroom/content/2007/10/content/printable/RiceReport10-2007.pdf. Accessed 14 May 2014

USDA (2013) Release No. 0127.13: Statement on the Detection of Genetically Engineered Wheat in Oregon. http://www.usda.gov/wps/portal/usda/usdahome?contentid=2013/06/0127.xml. Accessed 19 September 2014

USDA Economic Research Service (2014) Adoption of genetically engineered crops in the US. http://www.ers.usda.gov/publications/err-economic-research-report/err162.aspx. Accessed 14 May 2014

Watrud LS, Lee EH, Fairbrother A, Burdick C, Reichman JR, Bollman M, Storm M, King GJ, Van de Water PK (2004) Evidence for landscape-level, pollen mediated gene flow from genetically modified creeping bentgrass with CP4 EPSPS as a marker. Proc Natl Acad Sci U S A 101:14533–14538

Weiss R (2006) Firm Blames Farmers, 'Act of God' for Rice Contamination. Washington Post, USA. http://www.washingtonpost.com/wp-dyn/content/article/2006/11/21/AR2006112101265.html. Accessed 14 May 2014

Westphal SP (2001) Pig out. New Sci 2301:14. http://www.newscientist.com/article/mg17123011.700-pig-out.html#.U3Ec969OX5o. Accessed 14 May 2014

Yu-Tzu C (2003) Concern Over GM Papayas Raised by Jao. Taipei Times, Taiwan. http://www.taipeitimes.com/News/taiwan/archives/2003/09/16/2003068021. Accessed 14 May 2014

Zapiola ML, Mallory-Smith CA (2012) Crossing the divide: gene flow produces intergeneric hybrid in feral transgenic creeping bentgrass population. Mol Ecol 21:4672–4680

Zi X (2005) GM rice forges ahead in China amid concerns of illegal planting. Nat Biotechnol 23:637

12

Prevalence of *Campylobacter jejuni* in chicken produced by major poultry companies in Saudi Arabia

Hany M Yehia[1,2*] and Mosffer M AL-Dagal[1*]

Abstract

Background: Campylobacter is a foodborne pathogen that is commonly associated with chicken. The aim of this work was to evaluate the prevalence of *Campylobacter jejuni* (as affected by refrigerated storage) in chicken samples obtained from the wholesale poultry market in the northern part of Riyadh City, Saudi Arabia.

Findings: A gradual increase in the number of positive samples was noted during storage at 4°C. On days 1, 3, and 7, the number of positive samples were 10 (30.305%), 15 (45.45%), and 27 (81.81%), respectively. Of 99 tested samples, 52 (52.25%) were positive for *Campylobacter jejuni*. Protein profiling by Sodium dodecyl sulfate -Polyacrylamide gel electrophoresis (SDS-PAGE) was used to identify *Campylobacter jejuni*. The results were verified using Analytical Profile Index (API Campy system, Marcy l'Etoile, France). Forty-three (82.69%) positive isolates were identified as *C. jejuni subsp. jejuni* 2, 5 isolates as *C. jejuni subsp. jejuni* 1 (9.61%), and 4 isolates as *C. jejuni subsp.* doylei (7.69).

Conclusion: *C. jejuni* positive samples increased rapidly during storage at 4°C for approximately 1 wk. Our results also indicated a connection between the protein profiles on SDS-PAGE and API Campy used for the identification of *C. jejuni*.

Keywords: *Campylobacter jejuni*; Chicken; Refrigeration storage

Introduction

Campylobacter jejuni is a well-known food-borne pathogen, transmitted to humans by the eating of warm-blooded animal meat, especially poultry (Jay et al. 2005; Nielsen et al. 1997). This bacterium is a major cause of food-borne diarrhea in many countries (Crushell et al. 2004; Iovine et al. 2008. It has gained more attention in the last 30 years because it has been recognized as a major cause of human illnesses, ranging from gastroenteritis to Guillain-Barré Syndrome (Khanna et al. 1996; Tauxe, 2001; Moore et al. 2005). The 2 most frequently occurring *Campylobacter* species that are of clinical significance because of meat consumption and meat products are *C. jejuni* and *C. coli*. *Campylobacter jejuni* accounts for more than 90% of incidences of human campylobacteriosis (Lindmark et al. 2009). Campylobacteriosis in humans results from eating undercooked meat and/or contaminated meals (Corry and

Atabay 2001). Survey studies have revealed a high prevalence of *Campylobacter* in poultry meats (Dickins et al. 2002; Ridsdale et al. 1998; Stoyanchev et al. 2007).

Few studies on *Campylobacter* in the Saudi Arabian food market have been performed. Therefore, this study investigates the prevalence of *Campylobacter jejuni* in locally produced refrigerated chicken carcasses, as affected by storage time. Whole-cell protein profiles of presumptive Campylobacter isolates were compared with the standard strain of *Campylobacter jejuni* ATCC 33291 on SDS page. High degree of similarity within standard strain was confirmed by biochemical identification using an API CAMPY biotyping identification system.

Materials and methods
Sample collection

Whole chicken carcasses (n = 99) were obtained from a wholesale poultry market located in the northern part of Riyadh City, Saudi Arabia. The samples were collected from 11 major national poultry companies (designated by the letters A through K). Nine samples were collected from each company. The samples were transported at

* Correspondence: hanyyehia43@yahoo.de; maldagal@ksu.edu.sa
[1]Food Science and Nutrition Department, College of Food and Agricultural Sciences, King Saud University, Riyadh, Saudi Arabia
Full list of author information is available at the end of the article

refrigeration temperature to the Food Microbiology Laboratory, College of Food and Agricultural Sciences, King Saud University.

Experimental design

Refrigerated chicken samples were divided into 3 groups. The first group (including 33 samples, 3 samples (**3 runs**) per company) was tested for the presence of *Campylobacter jejuni* on Day 1 (the purchase date). Similarly, the next 2 groups (33 samples each) were tested for the micro-organism at Day 3 and Day 7 after the purchase date and stored in refrigerator at 4°C.

Isolation and identification of *Campylobacter*

Campylobacter jejuni was isolated according to the methods described by the Food Safety and Inspection Service (FSIS) (1998). Each carcass was rinsed in a sterile plastic bag with the addition of 200 mL of 0.1% peptone by manual shaking for 60 s. Ten milliliters of the rinse was centrifuged at 5000 rpm for 5 min, and 2 loops of the pellet were streaked on modified charcoal cefo-perazone deoxycholate agar (mCCDA, Oxoid CM739, Basingstoke, Hampshire, UK) and *Campylobacter*-selective agar (Preston, Oxoid CM0689). Preston Campylobacter selective supplement (Oxoid, SR0117) and lysed horse blood (Oxoid, SR0048) were used for the selective isola-tion of *Campylobacter jejuni* from the samples (Figueroa et al. 2009). As a confirmation step, 10 mL of each rinse fluid was transferred to 90 mL of Preston enrichment broth (to prepare Preston Campylobacter selective enrich-ment broth, 12.5 g of Nutrient Broth No.2 (Oxoid CM0067) was dissolved in 475 ml of distilled water and sterilized by autoclaving at 121°C for 15 minutes. Twenty five 25 ml of Lysed Horse Blood (Oxoid SR0048), 1 vial of Preston Campylobacter Selective Supplement (Oxoid SR0117) and 1 vial of Campylobacter Growth Supplement (Oxoid, SR0232) were added to the cooled 475 ml medium. Five ml volumes were aseptically dispensed in sterile small screw-capped bottles and incubated at 37°C for 48 h in a gas mixture of BBL GasPak, 70304 (Becton Dickinson and Cockeysville, MD, USA), (5% O_2, 10% CO_2, and 85% N_2). The enrichment was streaked onto selective media, and the plates were incubated at 42°C for up to 48 h under microaerophilic conditions. After incubation, the plates were inspected for presumptive colonies before Gram staining, and cells resembling with *Campylobacter* were subcultured onto mCCDA by streaking colony method and incubated for 2 to 5 d at 42°C under microaerophilic conditions.

Polyacrylamide Gel Electrophoresis (PAGE)

The preparation of isolates for SDS-PAGE and the running of the samples were performed according to the method by Scarcelli et al. 2001. Electrophoresis was performed in a 12% polyacrylamide running gel and a 4% stacking gel, with a 0.025 M Tris 0.19 M glycine buffer pH 8.3, and 100 µL of a sucrose buffer (50 mM Tris–HCl, pH 8; 40 mM EDTA, pH 8; 0.75 M sucrose).

Preparation of cell extract

An overnight culture (100 µL) was inoculated into a 10 ml of fresh medium (Brain heart infusion-Oxoid, CM1135) and grown to an Optical Density (OD) 620 of 0.6 to 0.8 (3 to 4 h). The cells were collected and weighed, and 250 mg of cells were then suspended in 100 µL of a TES buffer (50 mM tris HCl, pH 8, 1 mM EDTA, 25% sucrose). Twenty microliters of lysozyme (50 mg/mL) and 5 µL mutanolysin (5000 u/mL) were added to the suspended cells in the TES buffer and incubated at 37°C for 30 min. Five to ten microliters of 20% SDS were added, and the contents were mixed until they became clear visible. The contents were stored at –20°C for 1 to 2 d (Ismail 2007).

Fifty-microliter extracts (standard and isolated bacteria) were loaded on SDS-PAGE. Electrophoresis was performed at 25°C in a vertical tank apparatus using a constant-voltage power supply, until a bromophenol blue tracking dye reached the bottom of the gel. Gels were stained with 0.25% Coomassie Brilliant Blue R-250 (Bio-Rad, Marnes-la-Coquette, France) in water: methanol: acetic acid (6.5:2.5:1) for 18 h at room temperature. Gel destaining was per-formed by continuous agitation in a methanol: acetic acid: water (20:10:70 v/v/v) solvent until obvious bands of pro-teins were obtained.

Whole-cell protein profiles of presumptive *Campylobacter* isolates were compared with the standard strain of *Campylobacter jejuni* ATCC 33291 on SDS-PAGE. A high degree of similarity with the standard strains was confirmed by biochemical identification using an API CAMPY bio-typing system and a catalase test.

Biochemical identification of *Campylobacter* isolates

BioMérieux API CAMPY was used according to the manu-facturer's instructions for the biochemical identification of *Campylobacter*. The BioMérieux API CAMPY strip con-sisted of 20 microtubes containing dehydrated substances, with each microtubule corresponding to an individual test. The 20 tests were divided into 2 parts; the first part was composed of enzymatic and conventional tests, and the second part comprised assimilation or inhibition tests (Gorman and Adley 2005). The results of the enzymatic tests were obtained with the addition of conventional reagents after 24 h of incubation at 37°C under aerobic conditions. The results of the assimilation and inhibition tests were recorded after 24 h at 37°C under microaero-philic conditions. Incubation was extended to 48 h if the succinate assimilation test was negative, as indicated by the manufacturer's instructions.

Results and discussion

Table 1 shows the prevalence of *C. jejuni* in refrigerated chicken samples, as affected by storage time. According to the table, the samples from all 11 poultry companies contained *C. jejuni* at one or more of the tested shelf-life dates. At Day 1 of the shelf life, *C. jejuni* was recovered from 30.30% of the tested samples (10 of 33 samples). This percentage increased to 44.45% and to 81.81%, after refrigeration for 3 d and 7 d, respectively, indicating the positive effect of natural enrichment. Poultry companies B, C, I, and J had included highly positive samples for *C. jejuni* (n = 6; 66.7%), followed by A and E (n = 5; 55.6%), and then D, F, and K (n = 4; 44.4%). Only one sample (n = 1; 11.1%), from company G, was positive.

Survival of *C. jejuni* at refrigeration temperature approved in one study. The outcome of this work agreed with other research works indicating that the survival of *C. jejuni* at refrigeration temperature range (4 to 7°C) is better than range (20 to 30°C), and that it exhibits greater survivability at chilled temperatures (Karenlampi and Hanninen 2004; El-Shibiny et al. 2009). Rollins and Colwell (1986) showed that, at 4°C, *C. jejuni* can survive and remain at a viable but non-culturable stage for approximately 4 mo. Zhao et al. (2000) showed that *C. jejuni* survive for days or weeks in refrigerated foodstuffs. Vashin and Stoyanchev (2011) established that the microorganisms did not grow in chilled or frozen meat, but are able to survive during the storage period at 1-4°C. Campylobacter were detected up to the 25th days while at -18 to –20°C: up to the 45th day. The juice released into the bags from poultry liquidation is highly nutritive and forms microaerophilic conditions suitable for *Campylobacter*. Birk et al. (2004) confirmed that compounds present in chicken juice protect, and thus, prolong the survival of *C. jejuni* during storage at refrigerator temperatures. They confirmed that incubation at 5°C extended viability of cells of *C. jejuni*, and incubation at 10°C significantly prolonged the viability of *Campylobacter*. In addition, they found that storage in chicken juice at both 5°C and 10°C significantly prolonged the viable cells of *C. jejuni* compared to incubation in reference media. The total number of samples tested positive for *C. jejuni* was 52 (52.52%). Similar results in a previous survey study (Rahimi and Tajbakhsh 2008) found 56.1% of chicken samples to be positive for *Campylobacter*. Few studies related to the occurrence of Campylobacter in Saudi Arabian food market, from 2002 to 2004, two studies were conducted to assess specimens obtained from slaughter-houses in Bahrain and Saudi Arabia for *Campylobacter* contamination. In one study, specimens consisting of 35 whole chickens, 27 chicken livers, and 38 chicken faeces were assessed using a combination of three culture methods, and just over half (57%) were found to be positive for *Campylobacter* contamination (Ghazwan 2006). In another study, 60 chicken faeces specimens were assessed using a newly developed multiplex PCR technique with 100% *Campylobacter* detection (Al Amri et al. 2007). However, in both studies, *C. jejuni* accounted for the majority of *Campylobacter* detected. The findings of this study, which is the first of its kind in our setting, indicates a need for

Table 1 Prevalence of *Campylobacter jejuni* in refrigerated whole chicken carcasses of 11 poultry companies in Riyadh City wholesale market

Poultry samples	Group 1 First day of the production date			Group 2 3rd days of the production date			Group 3 7th day of the production date			Total Positive	%
	Run1	Run2	Run3	Run1	Run2	Run3	Run1	Run2	Run3	(out of 9)	
A	+	-	-	+	-	-	+	+	+	5	55.6
B	-	-	+	+	-	+	+	+	+	6	66.7
C	-	+	-	+	+	+	+	+	-	6	66.7
D	-	+	-	-	-	-	+	+	+	4	44.4
E	+	-	-	+	-	+	-	+	+	5	55.6
F	-	-	+	+	-	+	-	+	-	4	44.4
G	-	-	-	-	-	-	-	+	-	1	11.1
H	+	-	-	+	-	-	+	+	+	5	55.6
I	-	+	-	+	-	+	+	+	+	6	66.7
J	-	-	+	+	-	+	+	+	+	6	66.7
K	-	+	-	-	-	-	+	+	+	4	44.4
Number of total samples	33			33			33			99	
Number of positive sample	10			15			27			52	-
%	30.30			45.45			81.81			52.52	-

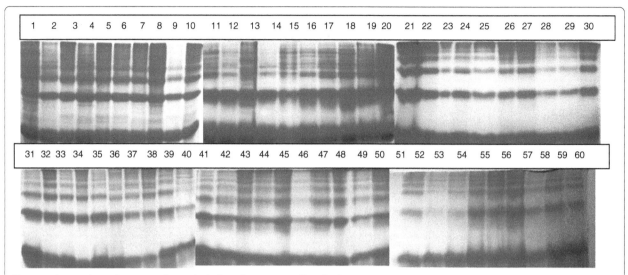

Figure 1 Comparison of patterns from of whole-cell protein profile of references strain *C. jejuni* ATCC 33291 (lane 1), and isolates from poultry carcass (lanes 2–60) on 12% SDS-PAGE.

increased surveillance and *Campylobacter* screening in food safety control to better protect consumers.

Figure 1 shows the protein profiles of the isolates as they appeared on the media of mCCDA and Preston agar compared with standard strains of *C. jejuni* ATCC 33291 on SDS-PAGE. The results demonstrated the presence of the common heavy protein band for 52 isolates compared with *C. jejuni* ATCC 33291. The whole-cell protein profiles of isolates were closely related to *C. jejuni* ATCC 33291. Some isolates could not be analyzed because they did not yield sufficient proteins after extraction, or the band did not resemble *C. jejuni* ATCC 33291. Any isolates showing discrepancy with the protein profile were excluded from confirmation of identification by using API CAMPY. Massai et al. (2007) explained that the whole-cell protein profile determined by SDS-PAGE expresses an important proportion of the genome. However, this expression may be modified by various factors. Therefore, this technique must be carefully controlled and standardized to obtain reproducible results. Advantages of the SDS-PAGE patterns are that they can be obtained in a short time, are reproducible, and do not require any sophisticated and expensive reagents or equipment compared with other molecular biology.

The polyacrylamide gel electrophoresis (PAGE) of bacterial proteins is an efficient technique for the classification of microorganisms, based on phenotypical characteristics expressed by their protein profiles (Scarcelli et al. 2001). Protein electrophoretic analysis of *Campylobacter spp.* in a polyacrylamide gel in the presence of SDS supplies data that can be used in epidemiological and taxonomic studies, as well as the identification of species and specific virulence factors (Dunn et al. 1987).

API campy strip system

After 24 to 48 h of incubation, the positive isolates obtained from SDS-PAGE were confirmed using API Campy, with 52 of 99 samples (52.52%) testing positive for both the first part and second part of the tests. Eleven of 99 (11.11%) isolates were negative. Forty-three of the 52 (82.69%) positive isolates were identified in both parts of the tests as *C. jejuni subsp. jejuni* 2. Five isolates were identified as *C. jejuni subsp. jejuni* 1 (9.61%), and 4 isolates were identified as *Campylobacter jejuni subsp. doylei* (7.96%). The reference strains were correctly identified by both systems as *C. jejuni ssp. jejuni* 1. Eleven isolates yielded discrepant identifications, and yielded profile codes absent in the API Campy database; therefore, they were considered negative results. Huysmans and Turnidge (1997) indicated that the correlation between API Campy and the conventional tests was 100% for the identification of *C. jejuni*.

Conclusion

The results show that *Campylobacter jejuni* was prevalent in poultry meat samples collected from wholesale markets in Riyadh City, Saudi Arabia. *C. jejuni* positive samples increased rapidly during storage at 4°C for approximately 1 wk. Our results also indicated a connection between the protein profiles on SDS-PAGE and API Campy used for the identification of *C. jejuni*.

Competing interests
The authors declare that they have no competing interests.

Authors' contributions
Conceived and designed the experiments: YHM and AMM. Performed the experiments: YHM . Analyzed the data: YHM and AMM. Wrote the manuscript: YHM and AMM. All authors read and approve the final manuscript.

Author details
[1]Food Science and Nutrition Department, College of Food and Agricultural Sciences, King Saud University, Riyadh, Saudi Arabia. [2]Food Science and Nutrition Department, Faculty of Home Economics, Helwan University, Cairo, Egypt.

References

Al Amri A, Senok AC, Ismaeel AY, Al-Mahmeed AE, Botta GA (2007) Multiplex PCR for direct identification of *Campylobacter* spp. in human and chicken stools. J Med Microbiol 56:1350–1355

Birk T, Ingmer H, Andersen MT, Jorgensen K, Brondsted L (2004) Chicken juice, a food-based model system suitable to study survival of *Campylobacter jejuni*. Lett Appl Microbiol 38:66–71

Corry JE, Atabay HI (2001) Poultry as a source of Campylobacter and related organisms. Symp Ser Soc Appl Microbiol 30:96S–114S

Crushell E, Harty S, Sharif F, Bourke B (2004) Enteric Campylobacter: purging its secrets? Pediatr Res 55:3–12

Dickins MA, Franklin S, Stefanova R, Schutze GE, Eisenach KD, Wesley I, Cave MD (2002) Diversity of Campylobacter isolates from retail poultry carcasses and from humans as demonstrated by pulsed-field gel electrophoresis. J Food Prot 65:957–962

Dunn BE, Blaser MJ, Snyder EL (1987) Two dimensional gel electrophoresis and immunoblotting of *Campylobacter* outer membrane proteins. Infect Immun 55:1564–1572

El-Shibiny A, Scott A, Timms A, Metawea Y, Connerton P, Connerton I (2009) Application of a group II Campylobacter bacteriophage to reduce strains of *Campylobacter jejuni* and *Campylobacter coli* colonizing broiler chickens. J Food Prot 72:733–740

Figueroa G, Troncoso M, Lopez C, Rivas P, Toro M (2009) Occurrence and enumeration of *Campylobacter* spp. during the processing of Chilean broilers. BMC Microbiol 9:94

Food Safety and Inspection Service (FSIS) (1998) Isolation, Identification and Enumeration of *Campylobacter jejuni/coli* from meat and poultry products. Microbiology Laboratory Guidebook. Chapter 3rd edition, United States Department of Agriculture, Washington D.C.

Ghazwan J (2006) Public Health Importance of *Campylobacter jejuni* in Poultry [dissertation]. Arabian Gulf University, Manama, Bahrain, p 66

Gorman R, Adley CC (2005) Campylobacter Isolation, Identification, and Preservation .From: Methods in Biotechnology, Vol. 21. In: Adley CC (ed) Food Borne Pathogens: Methods and Protocols. © Humana Press Inc, Totowa, NJ

Huysmans MB, Turnidge JD (1997) Disc susceptibility testing for thermophilic Campylobacters. Pathology 29:209–216

Iovine NM, Pursnani S, Voldman A, Wasserman G, Blaser MJ, Weinrauch Y (2008) Reactive nitrogen species contribute to innate host defense against *Campylobacter jejuni*. Infect Immun 76:986–993

Ismail EA (2007) Ph. Thesis Characterization and Genetic Improvement of Lactobacillus for Application in Probiotics Dairy Probiotic Dairy Products. University of Kiel, Germany

Jay J, Loessner M, Golden DA (2005) Modern Food Microbiology, 7th edition. Springer, New York

Karenlampi R, Hanninen ML (2004) Survival of *Campylobacter jejuni* on various fresh produce. Int J Food Microbiol 97:187–195

Khanna PN, Kumar A, Singh AK, Khan IA (1996) Thermophilic Campylobacters - public health importance and our observations. Indian J Comp Microbiol Immunol Infect Dis 17:32–40

Lindmark H, Boqvist S, Ljungstrom M, Agren P, Bjorkholm B, Engstrand L (2009) Risk factors for Campylobacteriosis: an epidemiological surveillance study of patients and retail poultry. J Clin Microbiol 47(8):2616–2619

Massai R, Bantarc C, Lopardo H, Vay C, Gutkind G (2007) Whole-cell protein profiles are useful for distinguishing enterococcal species recovered from clinical specimens. Rev Argent Microbiol 39:199–203

Moore JE, Corcoran D, Dooley JSG, Fanning S, Lucey B, Matsuda M, McDowell R, O'Riordan L, O'Rourke M, Rao JR, Rooney PJ, Sails A, Whyte P (2005) Campylobacter. Vet Res 36:351–382

Nielsen EM, Engberg J, Madsen M (1997) Distribution of serotypes of *Campylobacter jejuni* and *C. coli* from Danish patients, poultry, cattle and swine. FEMS Immunol Med Microbiol 19:47–56

Rahimi E, Tajbakhsh E (2008) Prevalence of Campylobacter species in poultry meat in the Esfahan city, Iran. Bulg J Vet Med 11:257–262

Ridsdale JA, Atabay HI, Corry JEL (1998) Prevalence of Campylobacters and Arcobacters in ducks at the abattoir. J Appl Microbiol 85:567–573

Rollins DM, Colwell RR (1986) Viable but non culturable stage of *Campylobacter jejuni* and its role in survival in the natural aquatic environment. Appl Environ Microbiol 52:531–538

Scarcelli E, Elizabeth OC, Margareth EG, Maristela VC, Erna EB, Ana PT (2001) Comparison of electrophoretic protein profiles of *Campylobacter jejuni subsp. Jejuni* isolated from different animal species. Braz J Microbiol 32(4):286–292

Stoyanchev T, Vashin I, Ring C, Atanassova V (2007) Prevalence of *Campylobacter* spp. in poultry and poultry products for sale on the Bulgarian retail market. Anton Leeuw 92:285–288

Tauxe RV (2001) Incidence, Trends and Sources of Campylobacteriosis in Developed Countries: An Overview. In the Increasing Incidence of Human Campylobacteriosis. Report and Proceedings of a WHO Constitution of Experts Copenhagen, Denmark, 21–25 November 2000, pp. 42–43. World Health Organization, Geneva

Vashin IT, Stoyanchev TT (2011) Influence of temperature on *Campylobacter jejuni* survival rates in pork meat. Bulg J Vet Med 14(1):25–30

Zhao T, Doyle MP, Berg DE (2000) Fate of *Campylobacter jejuni* in butter. J Food Prot 63:120–122

13

Preliminary evaluation of good sampling locations on a pig carcass for livestock-associated MRSA isolation

Marijke Verhegghe[1,2*], Lieve Herman[1], Freddy Haesebrouck[2], Patrick Butaye[2,3], Marc Heyndrickx[1,2] and Geertrui Rasschaert[1]

Abstract

Background: The presence of livestock-associated methicillin-resistant *Staphylococcus aureus* (LA-MRSA) in livestock animals, especially in pigs, gave rise to concerns of pork being a possible MRSA source to the human population. Monitoring the flow-through of LA-MRSA throughout the meat production chain could be useful. Here, the optimal sampling location for LA-MRSA isolation on pig carcasses was determined.

Findings: In one slaughterhouse, 40 cooled carcass halves from one LA-MRSA-positive herd were sampled on six carcass sites (ham, belly, back, forelimb, sternum and abdominal cavity). The obtained MRSA isolates were characterized using Pulsed Field Gel Electrophoresis. Without enrichment of the samples, no MRSA was isolated from the carcasses. After enrichment, MRSA was isolated from 19 out of 40 (47.5%) carcasses. The forelimb appeared to be the most contaminated part of the carcass (17/19 carcasses). Three pulsotypes were detected and the predominant pulsotype was also the herd pulsotype that was determined in our previous study.

Conclusions: The present study demonstrated that the forelimb is a good sampling location for LA-MRSA. For good determination of LA-MRSA on carcasses, enrichment is needed. Only LA-MRSA was isolated. Moreover, the farm strain was isolated from the carcasses, which indicates that transmission from the primary production throughout the slaughterhouse occurred. The results suggest that good hygiene practices in slaughterhouses are important to reduce the transmission of LA-MRSA to the human population.

Keywords: Livestock-associated MRSA; Carcasses; Slaughterhouse; Best sampling location

Findings

At present, livestock-associated methicillin-resistant *Staphylococcus aureus* (LA-MRSA) can be found in the majority of pigs, which could be a potential reservoir for the general human population (Vanderhaeghen *et al.*, 2010). Therefore, a possible exposure route for LA-MRSA is thought to be pork, although low numbers of LA-MRSA have been found indicating a low risk of exposure to the human population (Van Loo *et al.*, 2007; Weese *et al.*,

2010). Pig carcasses can be contaminated with LA-MRSA at the slaughterhouse with farm and/or slaughterhouse strains. For Hazard Analysis and Critical Control Points (HACCP) verification purposes and compliance with microbiological criteria for foodstuffs, control samplings for *Salmonella* occur on a two-weekly basis at the slaughterhouse (European Commission, 2005). To our knowledge, no guidelines are available for the detection of LA-MRSA on pig carcasses. In the present study, different sampling locations for LA-MRSA on a pig carcass were evaluated at the slaughterhouse based on the protocol for *Salmonella* testing for the compliance with microbiological criteria or foodstuffs (EC, 2005). Moreover, Pulsed Field Gel Electrophoresis (PFGE) was performed to gain insight into the genetic variety of, and possible sources for, the obtained isolates.

* Correspondence: marijke.verhegghe@ilvo.vlaanderen.be
[1]Institute for Agricultural and Fisheries Research (ILVO), Technology and Food Science Unit, Food safety research group, Brusselsesteenweg 370, Melle 9090, Belgium
[2]Department of Pathology, Bacteriology and Avian Diseases, Ghent University, Faculty of Veterinary Medicine, Salisburylaan 133, Merelbeke 9820, Belgium
Full list of author information is available at the end of the article

In 2012, sampling was performed in one slaughter-house, located in the northern part of Belgium. The slaughtered pigs originated from a LA-MRSA-positive farm (Verhegghe *et al.*, 2014). Prior to slaughter, the herd (consisting of 120 animals) was loaded onto a clean truck and transported immediately to the slaughter-house, where it was slaughtered first that day. Approximately two hours after slaughter, 40 carcasses out of 120 were randomly chosen and one half of each carcass was sampled in the cooling room: 19 right-carcass halves and 21 left-carcass halves. Each carcass half was sampled at six places, being the four locations as described by Ghafir *et al.* (2005) together with the belly and the back (Figure 1). The outside of the carcass half was sampled at the ham, the belly and the back and the inner part at the inner side of the forelimb, the sternum and the abdominal cavity. From each sampling location, 100 cm^2 was swabbed with an envirosponge (3 M DrySponge; BP133ES; Led Techno; St.-Paul, MN, US), premoistened with 7 ml salt-enriched (6.5%; Sodium chloride; 1.06404; Merk, Darmstadt, DE) Mueller Hinton Broth (MHB CM0405; Oxoid, Basingstoke, UK). All samples were transported and processed immediately upon arrival at the laboratory (two to three hours after the sampling event). Sixty-three ml salt-enriched MHB was added to the sponges; mixed manually for 30 seconds and 100 µl was spread-plated onto a chromogenic MRSA selective medium (Chrom-ID™ MRSA; BioMerieux, Marcy l'Etoile, FR) after which the plates were incubated overnight (18-20 h, 37°C). A ten-fold dilution series of this broth was made in salt-enriched MHB up to dilution 10^{-3}. The dilution series was

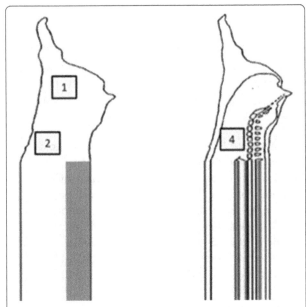

Figure 1 The different locations on the carcass (1: the ham, 2: the belly, 3: the back, 4: the abdominal cavity, 5: the sternum, 6: the inner part of the forelimb).

incubated overnight (37°C for 18-20 h), after which the enrichments were spread-plated and incubated onto Chrom-ID™ MRSA as described above. One suspect colony per plate was purified onto Chrom-ID™ MRSA and Tryptone Soy Agar (TSA;. Pure isolates were stored at −20°C in brain-heart infusion broth (BHI; CM0225; Oxoid, Basingstoke, UK) supplemented with glycerol (15% wt/vol; Fisher Scientific, Leicestershire, UK). From each isolate, DNA was extracted according to Strandén *et al.* (2003) and then stored at −20°C until further use. A MRSA-specific multiplex PCR identified MRSA isolates (Maes *et al.*, 2002). A carcass half was considered MRSA-positive if MRSA was isolated from at least one location. A chi-square test and a Fisher's exact test were used to analyze the sampling results of the carcass halves and the results of the different locations, respectively. Given that no positive samples were observed on the location "back", these data were excluded from the analysis. Analysis was performed in SPSS statistics 19 (IBM, Chicago, IL, US) and for all analyses, $P < 0.05$ was considered significant. A CC398-specific PCR, targeting the restriction-modification system encoding *sau*1*hsd*S1, and PFGE using *Bst*ZI (Promega, Madison, WI, US) as a restriction enzyme were performed with all obtained isolates as described by Stegger *et al.* (2011) and Rasschaert *et al.* (2009), respectively. The obtained PFGE profiles were analyzed with Bionumerics version 6.5 (Applied Maths, St.-Martens-Latem, BE) using the unweighted pair group method using averages (UPGMA) with the Dice coefficient (tolerance 1%, tolerance change 1% and optimization 1%). Pulsotypes were determined based on a delineation level of 97%. Comparison of the pulsotypes with the herd pulsotype was also performed in Bionumerics (Verhegghe *et al.*, 2014).

All MRSA isolates that were found, belonged to CC398, the LA-MRSA complex. LA-MRSA was not detected on the carcass halves after direct plating, indicating that −for this particular case- the level of LA-MRSA contamination on the carcass halves was lower than the detection limit (<7 cfu/cm^2). However, it is possible that an underestimation of the LA-MRSA presence occurred. Samples were taken after rapid cooling of the carcasses. *S. aureus* is able to bind strongly to corneocytes, which could imply that sponge swabbing might not be sufficient to collect all MRSA present and a more destructive method, such as cutting of slices is needed as was reported for *Salmonella* or *E. coli* (Ghafir and Daube, 2008; *Martinez et al.*, 2010). Another possibility could be the reduced viability of MRSA after cooling of the carcasses, even though *S. aureus* is able to persist colder temperatures (Onyango *et al.*, 2012).

After enrichment, LA-MRSA was isolated from 19 out of 40 carcass halves, i.e. 10 right halves and 9 left halves, which was not statistically significant (Chi-square, p = 0.88). From 17 out of 19 halves, LA-MRSA was isolated from only one site (15 times from the forelimb; once from the abdomen and intestine cavity). From two out of 19 halves LA-MRSA was isolated from two sites (forelimb-ham and forelimb-sternum). From 18 halves, MRSA was only detected in the initial enrichment broth, whereas on one half, MRSA was detected up to 10^{-2} dilution of the enrichment broth. The MRSA prevalence (47.5%) determined in our study, which investigated one herd, was higher than the prevalences observed in two other studies, being 6% and 4% of the carcasses (Beneke *et al.*, 2011 and Hawken *et al.*, 2013, respectively).

Most LA-MRSA isolates were found on the forelimb (17 carcass halves) and once on the ham, belly, abdominal cavity and sternum (Fisher's exact, P < 0.001). No forelimbs of the carcasses were sampled in the other studies mentioned above, but MRSA was predominantly found on the carcass shoulders, which is in close proximity to the lowest parts of the carcass. In the slaughterhouse, it was noticed that the bottom of the carcass, where the forelimb was located, was visually the dirtiest part of the carcass. In the evisceration room, the lower part of the carcasses came regularly in contact with the side of the evisceration platform. Since MRSA is able to survive the slaughter process and environmental contamination could also occur, the forelimb is an accurate sampling location for MRSA detection. However, the forelimb is not consumed very often in Belgium, which implies that this location is not appropriate to study MRSA transmission to the human population and other locations such as the back or the ham should be considered. However, only one slaughterhouse and carcass halves of only one MRSA-positive herd were sampled during the present study and only one sampling scheme was used. Therefore, to draw firmer conclusions, an additional study including samplings of pig carcasses of different herds at different slaughterhouses and using additional sampling sites is needed to confirm the forelimb as best sampling location for detection of LA-MRSA.

Twenty-two isolates were retrieved from 19 carcass halves and all isolates were identified as MRSA CC398 indicating that -in this case- no human strains had contaminated the carcasses during the slaughter process. Three pulsotypes were found after characterizing the isolates with PFGE. Pulsotype I was retrieved from 18 out of 22 isolates (17 carcass halves) (Figure 2). On one carcass half, pulsotypes I and II were isolated from the forelimb in different dilutions. Pulsotype I was also the only pulsotype found in the herd where the carcass halves originated from (Verhegghe et al., 2014; Figure 2). When a pig is colonized with LA-MRSA, this bacterium can be isolated from the skin and the nares (Broens *et al.*, 2012; Crombé *et al.*, 2012; Szabó *et al.*, 2012). It appears that LA-MRSA is not eliminated from the carcass during the slaughter process. For example, singeing of the carcass might be insufficient to eliminate bacteria from the lower part of the carcass, resulting in MRSA detection on the forelimbs. Besides the herd pulsotype, a minority of isolates with other pulsotypes were found on the carcasses. This indicates that various LA-MRSA strains might be widespread within the slaughterhouse environment, including tracks, lairage and the slaughterline. This could result in cross-contamination of the carcasses, since it has been reported that at the end of a slaughter day MRSA CC398 was widespread in the environment of pig and broiler slaughterhouses (Mulders *et al.*, 2010; Van Cleef *et al.*, 2010; Gilbert *et al.*, 2012). Nevertheless, further research is needed to investigate possible transmission routes for carcasses in a slaughterhouse. In addition, molecular characterization of the obtained isolates would be very valuable to better assess the transmission risk of (LA-)MRSA to the human population through contaminated carcasses and pork.

In conclusion, during the present study performed in one slaughterhouse, it was shown that LA-MRSA is regularly present on carcasses. After enrichment, the forelimb appeared a good sampling location to detect LA-MRSA on a carcass half. During the present study, only LA-MRSA was isolated. Moreover, the dominant pulsotype was isolated from the animals of the LA-

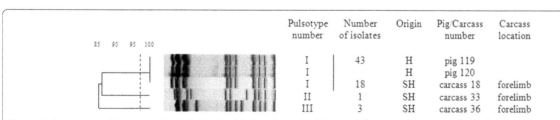

	Pulsotype number	Number of isolates	Origin	Pig/Carcass number	Carcass location
	I	43	H	pig 119	
	I		H	pig 120	
	I	18	SH	carcass 18	forelimb
	II	1	SH	carcass 33	forelimb
	III	3	SH	carcass 36	forelimb

Figure 2 Comparison of the three obtained slaughterhouse pulsotypes with the herd pulsotype (two out of 43 isolates). A delineation level of 97% (dotted line) was applied to discriminate the different genotypes. Consecutive the dendrogram, pulsotype pattern, pulsotype number and number of isolates belonging to the pulsotype on the total number of typed isolates are shown. For each example, the origin (H: herd, SH: slaughterhouse), pig/carcass number and carcass location is given.

MRSA-positive herd and from their carcasses, indicating a transmission from the primary production. The retrieval of other pulsotypes on the carcass halves implies that contamination of the carcasses from the slaughterhouse environment may also occur.

Abbreviations

LA-MRSA: Livestock-associated methicillin-resistant *Staphylococcus aureus*; CC: Clonal complex; ST: Sequence type; PFGE: Pulsed Field Gel electrophoresis; UPGMA: Unweighted pair cluster method using averages.

Competing interests

The authors declare that they have no competing interests.

Author's contributions

MV carried out the samplings, the sample analyses and drafted the manuscript. FH, PB, LH, MH and GR participated in the design of the study. GR coordinated the study and helped to draft the manuscript. All authors read and approved the final manuscript.

Acknowledgements

This research was funded by the Institute for the Promotion of Innovation by Science and Technology in Flanders (IWT) project 070596. We thank Pieter Siau for the sampling and laboratory assistance and the slaughterhouse for the collaboration to this study. Special thanks to Miriam Levenson for the English-language editing of this manuscript.

Author details

[1]Institute for Agricultural and Fisheries Research (ILVO), Technology and Food Science Unit, Food safety research group, Brusselsesteenweg 370, Melle 9090, Belgium. [2]Department of Pathology, Bacteriology and Avian Diseases, Ghent University, Faculty of Veterinary Medicine, Salisburylaan 133, Merelbeke 9820, Belgium. [3]Department of Bacteriology and Immunology, Veterinary and Agrochemical Research Centre (VAR), Groeselenberg 99, Brussels 1180, Belgium.

References

Beneke B, Klees S, Stührenberg B, Fetsch A, Kraushaar B, Tenhagen B-A (2011) Prevalence of methicillin-resistant *Staphylococcus aureus* in a fresh meat pork production chain. J Food Protect 74:126–129

Broens EM, Espinosa-Gongora C, Graat EAM, Vendrig N, Van de wolf PJ, Guardabassi L, Butaye P, de Jong MCM, Van de Giessen AW (2012) Longitudinal field study on transmission of MRSA CC398 within pig herds. BMC Vet Res 8: 58. doi:10.1186/1746-6148-8-58.

Crombé F, Vanderhaeghen W, Dewulf J, Hermans K, Haesebrouck F, Butaye P (2012) Colonization and transmission of methicillin-resistant *Staphylococcus aureus* ST398 in nursery piglets. Appl Environ Microb 78:1631–1634

European commission (2005) Commision regulation (EC) No 2073/2005 of 15 November 2005 on microbiological criteria for food stuffs (Text with EEA relevance) OJ L338 22.12.2005.

Ghafir Y, China B, Korsak N, Dierick K, Collard JM, Godard C, De Zutter L, Daube G (2005) Belgian surveillance plans to assess changes in *Salmonella* prevalence in meat at different production stages. J Food Protect 68:2269–2277

Ghafir Y, Daube G (2008) Comparison of swabbing and destructive methods for microbiological pig carcass sampling. Lett Appl Microbiol 47:322–326

Gilbert MJ, Bos MEH, Duim B, Urlings BAP, Heres L, Wagenaar JA, Heederik DJJ (2012) Livestock-associated MRSA ST398 carriage in pig slaughterhouse workers related to quantitative environmental exposure. Occup Environ Med 69:472–478

Hawken P, Weese JS, Friendschip R, Warriner K (2013) Longitudinal study of *Clostridium difficile* and methicillin-resistant *Staphylococcus aureus* associated with pigs from weaning through to the end of processing. J Food Protect 76:624–630

Maes N, Magdalena J, Rottiers S, De Gheldre Y, Struelens MJ (2002) Evaluation of a triplex PCR assay to discriminate *Staphylococcus aureus* from coagulase-negative staphylococci and determine methicillin resistance from blood cultures. J Clin Microb 40:1514–1517

Martinez B, Celda MF, Anastasio B, Garcia I, Lopez-Mendoza MC (2010) Microbiological sampling of carcasses by excision or swabbing with three types of sponge or gauze. J Food Protect 73:81–87

Mulders MN, Haenen APJ, Geenen PL, Vesseur PC, Poldervaart ES, Bosch T, Huijsdens XW, Hengeveld PD, Dam-Deisz WDC, Grant EAM, Mevius D, Voss A, Van de Giessen AW (2010) Prevalence of livestock-associated MRSA in broiler flocks and risk factors for slaughterhouse personnel in The Netherlands. Epidemiol Infect 138:743–755

Onyango LA, Dunstan RH, Gottfries J, Von Eiff C, Roberts TK (2012) Effect of low temperature on growth and ultra-structure of *Staphylococcus aureus*. PlosOne 7:e29031

Rasschaert G, Vanderhaeghen W, Dewaele I, Janez N, Huijsdens X, Butaye P, Heyndrickx M (2009) Comparison of fingerprinting methods for typing methicillin-resistant *Staphylococcus aureus* sequence type 398. J Clin Microb 47:3313–3322

Stegger M, Lindsay JA, Moodley A, Skov R, Broens EM, Guardabassi L (2011) Rapid PCR detection of *Staphylococcus aureus* clonal complex 398 by targeting the restriction-modification system carrying *sau*1-*hsd*S1. J Clin Microb 49:732–734

Strandén A, Frei R, Widmer AF (2003) Molecular typing of methicillin-resistant *Staphylococcus aureus*: Can PCR replace pulsed-field gel electrophoresis? J Clin Microb 41:3181–3186

Szabó I, Beck B, Friese A, Fetsch A, Tenhagen BA, Roesler U (2012) Colonization kinetics of different methicillin-resistant *Staphylococcus aureus* sequence types in pigs and host susceptibilities. Appl Environ Microb 78:541–548

Van Cleef BAGL, Broens EM, Voss A, Huijsdens XW, Zuchner L, Van Benthem BHB, Kluytmans JAJW, Mulders MN, Van de Giessen AW (2010) High prevalence of nasal MRSA carriage in slaughterhouse workers in contact with live pigs in The Netherlands. Epidemiol Infect 138:756–763

Van Loo IHM, Diederen BMW, Savelkoul PHM, Woudenberg JHC, Roosendaal R, Van Belkum A, Toom NLD, Verhulst C, Van Keulen PHJ, Kluytmans JAJW (2007) Methicillin-resistant *Staphylococcus aureus* in meat products, the Netherlands. Emerg Infect Dis 13:1753–1755

Vanderhaeghen W, Hermans K, Haesebrouck F, Butaye P (2010) Methicillin-resistant *Staphylococcus aureus* (MRSA) in food production animals. Epidemiol Infect 138:606–625

Verhegghe M, Pletinckx LJ, Crombé F, Van Weyenberg S, Haesebrouck F, Butaye P, Heyndrickx M, Rasschaert G (2014) Genetic diversity of livestock-associated MRSA isolates obtained from piglets from farrowing until slaughter age on four farrow-to-finish farms. Vet Res 45:89, 10.1186/s13567-014-0089-4

Weese JS, Reid-Smith RJ, Rousseau J, Avery B (2010) Methicillin-resistant *Staphylococcus aureus* (MRSA) contamination of retail pork. Can Vet J 51:749–752

14

Inorganic contaminants and composition analysis of commercial feed grade mineral compounds available in Costa Rica

Fabio Granados-Chinchilla[1*], Sugey Prado Mena[1] and Lisbeth Mata Arias[1,2]

Abstract

Background: Heavy metals such as arsenic (As), mercury (Hg), lead (Pb) and cadmium (Cd), are potential toxic substances that may incorporate in productive systems in multiple ways including contaminated feedstuff. In this regard, we hypothesize that the main input of heavy metal contamination include mineral feed ingredients which, in turn, are included in compound feed to meet animals' nutritional requirements. Hence, we offer a comprehensive heavy metal determination on imported feed grade mineral supplement samples ($n = 435$), comprised of 27 different sources including calcium/phosphorus, iron, cobalt, copper, cobalt, manganese, iodide, sulfur, potassium, sodium, selenium and magnesium were collected from eight different local feedingstuff manufactures, mineral and heavy metal as cadmium (Cd), lead (Pb), arsenic (As) and mercury (Hg) analyses were performed using atomic absorption spectroscopy and microwave assisted digestion. Based on this premise, the main goals of the study were to determine mineral and trace mineral content and contrast these values with those advertised by the manufacturers and to determine heavy metal concentrations and compare these levels with the current regulation in commercially available mineral sources which are used in premixes and downstream formulation of compound feeds; a matter which remains undocumented.

Results: Our results show that occasionally mineral values for these supplements were lower than those declared. Additionally, several samples contravene, in at least one heavy metal, current regulations; samples in this condition correspond to 0.5 ($n = 2$), 13.8 ($n = 60$), 4.1 ($n = 18$) and 2.5 % ($n = 11$) for As, Hg, Pb and Cd, respectively with mercury as the most frequent mineral to surpass, in the majority of cases, current thresholds. Overall, 21.1 % ($n = 92$) of the samples exhibited concentrations of heavy metals above those stipulated by European guidelines. Meanwhile potassium chloride, ($n = 17$), exhibited the lower overall concentrations of heavy metals.

Conclusion: Samples of mineral origin may surpass, in some cases with elevated concentrations, permitted levels of undesirable substances, therefore, a monitoring programme for mineral ingredients in our country is recommended.

Keywords: Heavy metals; Mineral ingredients; Animal feed; Guaranteed analysis

Background

Animal feed ingredients that constitute complete feed products are derived from a multitude of raw materials of plant and animal origin, as well as pharmaceutical and industrial sources. Many metallic compounds including calcium, copper, manganese, magnesium, and zinc compounds, as well as metal amino acid complexes (AAFCO 2014) are administered in animal feeds to satisfy nutritional needs of food-producing and companionship animals. Some minerals and/or premixes that may be by-products or co-products of industrial metal production might become contaminated with various heavy metals, (US FDA 2003) this is of particular concern from a public health perspective (Mamtani et al. 2011). In this regard, heavy metal contamination is especially important if one consider that a possible contamination in animal diets could consequently lead to transference to animal tissues and food products (Leeman et al. 2007).

* Correspondence: fabio.granados@ucr.ac.cr
[1]Centro de Investigación en Nutrición Animal (CINA), Universidad de Costa Rica, Ciudad Universitaria Rodrigo Facio, San José 11501-2060, Costa Rica
Full list of author information is available at the end of the article

Arsenic (As), cadmium (Cd), mercury (Hg), and lead (Pb) are widely dispersed in the environment. These elements have no known beneficial effects in humans or animals nor homeostasis mechanism for them (Poirier and Littlefield 1996). Furthermore, these metals are of special interest due to their inherent ability to bioaccumulate or biomagnify (Demirezen and Uruç 2006), transfer into and between food chains (Makridis et al. 2012) and their lack of biological functions (Singh et al. 2011). Although toxicity and the resulting threat to animal or human health of any contaminant are, of course, a function of concentration and inclusion rates in diet, it is well-known that sublethal chronic exposure to As, Cd, Hg, and Pb at relatively low levels can cause detrimental effects on physiological and/or biochemical processes (Sharma and Agrawal 2005).

AAFCO enlists different minerals used as feed ingredients and registers representative levels of toxic contaminants such as cadmium, arsenic, lead, and mercury in mineral feed ingredients, at the same time suggests guidelines for contaminants (based on toxicological capabilities) in complete feed and in mineral feed ingredients (AAFCO 2014).

Despite feedingstuff and its individual components (even those considered as minor ingredients) are the most probable and important way of acquiring these metals by husbandry animals, heavy metals may also enter productions systems in a variety of ways; these include atmospheric deposition, land application of inorganic fertilizers, biosolids, agrochemicals, and animal manure (Nicholson et al. 2003). However, the magnitude of these direct inputs will be determined by many indirect factors, such as farm location and type of production system (i.e. extensive or intensive) (López-Alonso et al. 2002; Blanco-Penedo et al. 2009).

It has been established that the most important source of heavy metal in concentrates and compound feeds are mineral supplements and premixes (Li et al. 2005) that are blended as additives during their formulation to supply the animal their dietary minimal intake of essential nutrients. However, it is still difficult to assess concentrations of toxic metal content in animal feed materials because of limited information, as most of the available data correspond to compound feed and few individual constituents whereas data on the particular mineral constituents to our knowledge is not existent. Hence, we found pertinent a comprehensive analysis of heavy metals in these mineral ingredients.

It is widely recognized that industrial and/or agricultural development is largely responsible for environmental pollution with toxic metals, though natural geochemical contamination has not been unheard of (Järup 2003). Most of these mineral sources are attained by industrial mining —mostly from metamorphic, igneous, volcanic, quarry or sedimentary rock (Elzea et al. 2006)— these types of raw materials often contain impurities, mainly fluorine (in the case of feed phosphates) and heavy metal compounds (e.g. Pb, Cd, As, Hg). Additionally, these products must be subjected to physicochemical treatments in order to meet the zootechnical requirements or transformed to soluble species with improved digestibility and assimilability (van Paemel et al. 2010).

As a country with non-mining, mineral processing or industrial capabilities Costa Rica imports these minerals from world mayor producers which include Netherlands, Peru, Chile, USA, China, Brazil, France among others (up to ca 1300 ton/550 thousand USD in 2014 alone) (PROCOMER 2014). Hence, the quality of feed produced in our country will depend on exporters to abide by current regulations. For example, the current all-European regulations in this field specify the permissible content of undesirable substances in animal feed (EC 2002; Official Journal of European Communities 2012).

Finally, due to a worldwide concern during feed production is the availability of supplies of raw ingredients (Van der Weijden et al. 2013) animal specialists must seek highly productive animals while assuring feed, feed supplies and food products (as well as the rest of the food chain) safety (EC 2000) but at the same time, using cost effective feedstock (e.g. technological processes by-products). Therefore, a decrease in exposure to undesirable contaminants through correct monitoring of these raw materials is paramount.

The aim of this study was to determine mineral, trace minerals and heavy metal contents in commercially available mineral sources which are used in premixes and downstream formulation of concentrates. The importance of these results are threefold as 1. They may play an important role as a standpoint to evaluate regionally the quality of raw materials as mineral supplements imported to the country 2. could help establish the regional and local guidelines for heavy metals maximum limits and 3. may serve as a guideline to manufacturers to make decisions with regard of the quality and safety of materials they use to prepare feeds.

Methods
Sampling

A total of 435 imported feed grade mineral supplement samples were selected for analysis. Samples were collected from eight plants that use these minerals in the country during 2012 and 2014 by a rolling census sampling method, taking into consideration the most common used ingredients for feed elaboration. A portion of the material was collected according to a defined sampling procedure that included only those containers which, when arriving to the manufacturing plant, were already in use. Trial units were obtained directly from container bags found in-plant and represented several batches. Samples were transported to the laboratory, a 100 g subsample of each mineral was homogenized, sieved and milled (up to a mm of particle size if deemed necessary) and kept in polyethylene sealed bottles at 24 °C until analyzed.

Mineral content

From 0.2 to 0.4 g sample was analyzed for total minerals resorting to wet mineralization carried out in an FOSS Tecator™ Digestor Auto (FOSS, Hillerod, Denmark) using a mixture of perchloric acid, nitric acid (Merck suprapur; E. Merck, Darmstadt, Germany) and water. Mineral species in the form of carbonates were digested *in situ* using a 6 mol/L solution of hydrochloric acid (Merck suprapur; E. Merck, Darmstadt, Germany) with mild heating in a hot plate at 300 °C, and analyzed using the flame atomic absorption spectrophotometry (FAAS) technique (Perkin-Elmer AAnalyst 800 atomic absorption spectrometer, Perkin-Elmer Corp., Norwalk, CT, USA). The radiation source, a Hollow Cathod Lamp for each metal, was used at wavelengths and spectral silt widths recommended by the manufacturer.

In the case of phosphorus, UV–vis spectrophotometry using molibdovanadate method was employed for measuring this mineral based on AOAC official method 935.13. Formation of the complex was obtained using 3 mL molybdovanadate solution instead of 2 mL as described in the original method.

Heavy metal analysis

Methods AOAC 986.15 and 999.10 were used during this survey. Briefly, 100 g subsample of each mineral was milled and from 0.2 to 0.4 g was analyzed for total heavy metals (Pb, Cd, As and Hg) by 90:10 nitric acid:hydrogen peroxide (both Merck suprapur; E. Merck, Darmstadt, Germany) microwave digestion using a Berghof Speedwave four (Harretstrasse, Eningen, Germany): Analyses were carried out either by flow injection metal hydride atomic absorption spectrophotometry (FI-MH-AAS, for Se and Hg) or graphite furnace atomic absorption spectrophotometry (GFAAS) technique (Perkin-Elmer AAnalyst 800 atomic absorption spectrometer, Perkin-Elmer Corp., Norwalk, CT, USA) in the case of As, Cd and Pb. Hydride generation was performed using a 30 g/L $NaBH_4$ in 10 g/L NaOH. The radiation source used was an Electrodeless Discharge Lamp for each metal used at wavelengths and a spectral slit widths recommended by the manufacturer.

Calibration curves

For metals analyzed with AAS or colorimetry, calibration curves were measure each time a metal was analyzed. Standards were prepared using commercially available NIST reference standard materials 10 mg/g (prepared from metal oxides and nitrates in HNO_3). All solutions were prepared with ultrapure water with a final conductivity of 0.055 µS/cm obtained using a Millipore Elix 35/

Milli-Q Advantage A10 system (Millipore, Bedford, MA, USA) immediately before use. Magnesium nitrate and ammonium dihydrogen phosphate in a concentration of 10 g/L were used as matrix modifiers during Cd, As and Pb graphite furnace analyses.

Sulfur analysis

Sulfur was determined gravimetrically after a microwave digestion using a concentrated nitric acid and hydrochloric acid mixture (65:37). Hereafter, a 10 mL aliquot of a 10 g/100 mL barium chloride solution is added to the resulting digest. After 12 hours of idle time at 4 °C, the resulting crystalline solid is retained on a previously weighted 542 Whatman filter paper (GE Healthcare Bio-Sciences, Pittsburg, USA), both filter and solid are dried afterwards. The recovered barium sulfate precipitate is weighted and calculated as sulfur in the original sample.

Anion analysis

Fluorine analysis, using an ion selective electrode, was performed to dicalcium and monocalcium phosphates. This assay was performed according to AOAC OMA 975.08. Briefly, a well-mixed test portion containing ca. 400 mg F^- was weighted into a, high-density polyethylene, 200 mL volumetric flask. A 20 mL aliquot 1 mol/L HCl was added and the mixture stirred during 20 min at high speed on magnetic stirrer. Solutions of sodium acetate and citrate are added to the mixture and the flask was made-up to volume with water. The mixture was transferred to a polyethylene sample cup. Total Fluorine concentration was determined by means of a calibration curve constructed with solutions of NaF (ACS grade, J.T. Baker, PA, USA) with concentrations between 0.3 to 10 mg/L Fluorine. Measurements were performed with a Fluorine sensitive electrode (Denver Instruments, Bohemia, NY, USA).

Chloride ion was determined by means of a potentiometric analysis according to AOAC OMA 969.10. Briefly, a test portion of ca. 0.25 grams was weighted in a 200 mL volumetric flask then 190 mL H_2O and 1 mL TISAB II were added. The flask was made-up to volume with water. This mixture was titrated with a 0.1 mol/L $AgNO_3$ solution up to inflection point Measurements were performed with a chloride sensitive electrode (Denver Instruments, Bohemia, NY, USA).

Iodine analysis was performed by the indirect determination of iodide ion equivalents using an iodine ISE electrode (6.0502.160, Metrohm, Ionenstrasse, Switzerland) in the case of metal iodides and etilendiamine diiodide ca. 70 mg and 200 mg, respectively of each sample were dissolved to a total volume of 50 mL in a volumetric flask with ultrapure water. For iodate and microgran™ samples 100–200 mg of each sample were dissolved in concentrated HNO_3 and mixed with 10 mL of a 100 mmol/L solution of hydrazine hydrate (ACROS Organics, Fair Lawn, NJ, USA), which oxidize IO_4^- ion to I_2, followed by a second reduction reaction from this species to I^- using a 100 mmol/L solution of sodium bisulfite to finally obtain a clear solution and finally adjusted pH of 7.0 before measurement. This procedure was adapted from an ISE-based method previously reported (Hasty 1973). Total iodide concentration was determined by a calibration curve constructed with solutions of KI (ACROS Organics, Fair Lawn, NJ, USA) with concentrations between 5.0×10^{-5} to 5×10^{-1} mol/L. Nitrate ion at 1 mol/L was added in all cases as an ionization buffer.

Carbonate ion was also analyzed potentiometrically according to AOAC OMA 955.01 using a glass pH electrode (Hanna Instruments, Carrolton, TX USA) and titrated using a 0.25 mol/L NaOH standard solution until sample solution attain pH 7. All anions were assayed using their respective electrodes coupled to a Denver Instruments 250 pH/ISE/potentiometric meter (Denver Instruments, Bohemia, NY, USA).

Quality control of results

An AAFCO Feed grade monocalcium phosphate sample (number 201298) was employed as a reference material throughout the heavy metal and Fluorine analyses. Beef cattle medicated feed sample (number 201228) was also used as a reference material for the rest of the minerals. Both samples were reprocessed and analyzed each time a given analyte was measured, in all cases z values $< |3|$ we considered, for our purposes, to be satisfactory. The methods stated above (including those used for sample pretreatment) are traditionally the most common analytical methods for the determination of total heavy metal residues in feed, feedstuffs and animal tissues. And have demonstrated to be sensitive enough to monitor compliance of statutory limits and are supported by law enforcing organizations (EC 2007). Limits of detection (established as S/N of 3.3, based on blank samples and regression curves) for As, Hg, Pb and Cd calculated as follow: 1000, 20, 600, 10 ng/kg, respectively.

Statistical analysis

ANOVA using Tukey *post hoc* was used to statistically compare metal content between mineral samples of different origin. In some specific cases, Dunnet's test was performed to demonstrate, for a specific element, if values obtained were significantly lower than those guaranteed by the manufacturer label. p values < 0.05 are considered to differ significantly.

Results and discussion

The typical concentrations of essential and contaminant heavy metals in common mineral ingredients used in feed manufacturing is offered. Some information regarding heavy metals in compound feed is presented as well.

From all the mineral sources analyzed, without exception, sulfate ion-containing minerals exhibit a significantly ($p < 0.001$) higher concentration of water (from 42.5 to 71.3 g/100 g dry matter) compared to other sources. This is an anticipated result as SO_4^{2-}, in accordance with Hofmeister series (dos Santos et al. 2010), is a small and multiply-charged ion, with high charge density, a poor disruptive of water's molecular binding (kosmotropic) (Wei et al. 2005) and thus exhibit stronger interactions with water (Plumridge et al. 2000). This said, is to be concluded that for all practical purposes in these type of salts, ions are to be treated as hydrated.

The case of calcium phosphates is singular. Degradation of orthophosphate into unavailable phosphate forms during manufacture (Hoffmann et al. 2011) is a possibility in those minerals which demonstrate lower values of phosphorus but calcium values within what is expected for these type of products. This occurs in exactly two ($n = 2$) samples of dicalcium phosphate (Table 1).

Due to the diverse chemical composition of both feed phosphate sources analyzed, large differences do exist in their phosphorus availability for different animal species (Fernandes et al. 1999; Viljoen 2001; Petersen et al. 2011). Phosphates from the same source and produced by the same process, using raw substances of similar quality should offer consistent phosphorus concentrations. In this respect, we suggest that chemical data should always be accompanied by bioavailability assays.

In a similar fashion, we have found, for the colorimetric assay of phosphorus, standard amounts of molybdovanadate to be inadequate after wet acid digestion, probably due an incomplete complexation of phosphorus by this reagent (data not shown). Therefore, we suggest using a stoichiometric excess of color-forming agent or another quantitative technique such as GFAAS or ICP-MS for total phosphorus quantitation.

Another distinctive feature of feed phosphates is that fluorine can be used as a quality control for the inorganic synthesis and inputs used during its manufacture (e.g. wet acid route). In this case, only 4.5 % ($n = 1/22$) of the samples, exceeded the 2 g/kg guideline for this anion (Table 1). So far, only dicalcium phosphate samples have exceeded regulatory guidelines for fluorine (Table 1). However, no significant differences ($p < 0.05$) were found for this nutrient when both types of phosphate sources were compared. Synthesis-wise new technologies for the production of feed phosphates have been introduced with the sole purpose of reducing this contaminant to acceptable levels (Hoffmann et al. 2011).

On another hand, in the case of macronutrients, a worrisome situation was evidenced by the detection of some inconsistencies in labeling (either by the mineral

Table 1 Mineral contents in typical mineral imported supplements employed in feedingstuff manufacture

Assay[a]	Mean	SD	Median	Max	Min	Guideline value or guaranteed analysis
Calcium and phosphorus-containing mineral sources						
Dicalcium phosphate, n = 22						
Dry matter, g/100 g	97.3	2.5	98.2	100.0	90.9	>88
Calcium, g/100 g[1; 4.5 %]	26.2	2.9	26.0	32.7	21.4	21.0 – 27.0
Phosphorus, g/100 g[1; 4.5 %]	19.5	1.7	19.8	22.0	15.5	18.0 – 21.0
Monocalcium phosphate, n = 22						
Dry matter, g/100 g	94.2	1.2	94.3	95.7	90.8	>88
Calcium, g/100 g[1; 4.5 %]	16.0	1.4	15.9	18.9	13.1	15.0 – 18.0
Phosphorus, g/100 g[1; 4.5 %]	21.0	1.2	20.9	23.6	18.0	21.0 – 23.0
Calcium carbonate, n = 21						
Dry matter, g/100 g	99.9	0.1	99.9	100.0	99.6	>88
Calcium, g/100 g[2; 9.5 %]	37.8	2.8	36.4	41.8	28.9	37.5 – 39.0
Carbonate (as $CaCO_3$), g/100 g	92.6	12.5	97.5	99.6	48.6	96.0 – 98.0
Iodine-containing mineral sources						
Calcium iodate, n = 7						
Dry matter, g/100 g	99.57	0.34	99.71	99.86	99.00	>88
Calcium, g/100 g	7.6	1.1	7.6	8.95	6.25	Not indicated
Iodide, g/100 g[1; 14.2 %]	61.4	4.7	62.6	64.7	50.3	63.5
Microgran™ I 10 % BMP (calcium iodide), n = 10						
Dry matter, g/100 g	99.45	0.24	99.51	99.76	98.90	>88.0
Calcium, g/100 g	16.9	5.1	18.4	20.8	4.8	-
Iodide, g/100 g[4; 40.0 %]	8.1	2.6	8.1	11.9	4.4	10.0
Etilendiamine diiodide, n = 7						
Dry matter, g/100 g	99.89	0.08	99.93	99.96	99.73	>88.0
Iodine, g/100 g[5; 83.3 %]	68.9	4.8	69.2	77.5	62.8	79.5
Potassium Iodide, n = 7						
Dry matter, g/100 g	99.7	0.1	99.7	99.4	99.8	>88.0
Potassium, g/100 g	21.1	1.9	19.8	24.5	19.5	-
Iodine, g/100 g[b]	68.9	5.5	70.1	74.9	62.5	68.0
Cobalt-containing mineral sources						
Cobalt carbonate, n = 13						
Dry matter, g/100 g	94.68	3.36	96.47	99.00	88.40	>88.0
Cobalt, g/100 g[7; 53.8 %]	49.4	10.7	45.4	82.4	39.7	46.0
Carbonate (as $CaCO_3$), g/100 g	62.4	14.4	68.8	84.1	43.9	Not indicated
Cobalt sulfate, n = 16						
Dry matter, g/100 g	67.83	9.77	64.50	98.37	63.20	>88.0
Cobalt, g/100 g[2; 12.5 %]	23.2	3.4	25.8	27.7	21.0	19.5-21.0
Microgran™ Co 5 % BMP (cobalt carbonate), n = 13						
Dry matter, g/100 g	97.9	0.6	97.7	100.0	97.3	>88.0
Carbonate (as $CaCO_3$), g/100 g	7.9	1.8	7.2	13.5	6.1	Not indicated
Cobalt, g/100 g[b]	6.3	1.4	5.7	10.7	4.8	5.0
Copper-containing mineral sources						
Copper oxide, n = 19						
Dry matter, g/100 g	99.9	0.2	99.8	100.0	99.1	>88.0

Table 1 Mineral contents in typical mineral imported supplements employed in feedingstuff manufacture *(Continued)*

Assay[a]	Mean	SD	Median	Max	Min	Guideline value or guaranteed analysis
Copper, g/100 g[3; 17.6 %]	71.7	5.9	71.3	85.0	59.6	75.0
Copper sulfate, n = 18						
Dry matter, g/100 g	73.4	6.5	71.9	99.7	68.8	>88.0
Copper, g/100 g[2; 12.5 %]	25.1	2.1	24.6	31.4	22.4	25.0
Sodium containing mineral sources						
Sodium chloride, n = 23						
Dry matter, g/100 g	96.2	6.1	98.8	99.9	71.7	>88.0
Sodium, g/100 g	36.6	4.0	36.4	46.7	26.7	38.5
Chloride (as NaCl), g/100 g[4; 19.0 %]	99.7	3.3	98.0	99.8	87.4	98.0
Iron-containing mineral sources						
Ferrous carbonate, n = 21						
Dry matter, g/100 g	99.0	0.54	98.9	100.0	98.15	>88.0
Iron, g/100 g[3; 15.8 %]	35.9	4.2	35.4	46.6	27.3	38.0
Carbonate (as $CaCO_3$), g/100 g	38.08	17.2	41.2	74.7	19.2	Not indicated
Ferric oxide, n = 16						
Dry matter, g/100 g	99.1	97.2	99.9	88.3	100.0	>88.0
Iron, g/100 g	37.6	13.0	37.3	65.0	16.2	Not indicated
Ferrous sulfate, n = 17						
Dry matter, g/100 g	75.4	17.3	62.6	99.1	62.0	>88.0
Iron, g/100 g[1; 6.3 %]	22.8	5.0	21.7	36.5	15.0	19.0
Magnesium-containing mineral sources						
Magnesium oxide, n = 21						
Dry matter, g/100 g	99.81	0.21	99.90	99.99	99.29	>88.0
Magnesium, g/100 g[6; 28.6 %]	42.1	15.1	40.5	85.4	22.4	50.0-54.0
Magnesium sulfate, n = 18						
Dry matter, g/100 g	63.1	9.2	60.8	99.8	57.5	>88.0
Magnesium, g/100 g[b]	8.8	1.1	8.6	11.7	6.8	6.8-11.6
Manganese-containing mineral sources						
Manganous oxide, n = 21						
Dry matter, g/100 g	99.6	0.24	99.7	99.9	98.9	>88.0
Manganese, g/100 g[7; 35.0 %]	58.6	6.6	56.2	77.8	50.4	60.0-62.0
Manganese sulfate, n = 15						
Dry matter, g/100 g	97.0	10.5	99.9	100.0	57.7	>88.0
Manganese, g/100 g[2; 13.3 %]	28.3	5.6	26.8	40.3	21.1	29.0-32.0
Sulfur-containing mineral sources						
Elemental sulfur, n = 20						
Dry matter, g/100 g	98.7	1.1	99.2	99.7	96.0	>88.0
Sulfur, g/100 g[4; 20.0 %]	98.4	1.7	99.0	99.6	93.7	99.5
Zinc-containing mineral sources						
Zinc oxide, n = 22						
Dry matter, g/100 g	99.5	1.7	99.9	100.0	91.5	>88.0
Zinc, g/100 g[1; 4.5 %]	76.0	12.3	72.4	108.4	58.3	72.0
Zinc sulfate, n = 18						
Dry matter, g/100 g	99.85	0.17	99.96	100.0	99.45	>88.0

Table 1 Mineral contents in typical mineral imported supplements employed in feedingstuff manufacture *(Continued)*

Assay[a]	Mean	SD	Median	Max	Min	Guideline value or guaranteed analysis
Zinc, g/100 g[8; 44.4 %]	32.4	4.1	32.0	43.2	25.1	35.0
Selenium-containing mineral sources						
Microgran™ Se 1 % BMP (sodium selenite), n = 13						
Dry matter, g/100 g	99.7	0.19	99.7	99.9	99.2	>88.0
Selenium, g/100 g	1.7	1.1	1.2	4.2	0.8	1.0
Sodium, g/100 g	0.63	0.12	0.61	0.93	0.49	Not indicated
Sodium selenite 1 %, n = 11						
Dry matter, g/100 g	99.52	0.37	99.64	99.95	98.65	>88.0
Selenium, g/100 g	1.36	0.77	1.14	3.58	0.68	1.0
Sodium, g/100 g	1.03	1.42	0.62	5.50	0.21	Not indicated
Sodium selenite, n = 8						
Dry matter, g/100 g	99.76	0.19	99.82	99.99	99.48	>88.0
Selenium, g/100 g	44.3	2.6	44.2	47.7	40.0	45-46
Sodium, g/100 g	24.1	3.25	24.6	29.3	19.0	26-27
Potassium-containing mineral sources						
Potassium Chloride, n = 17						
Dry matter, g/100 g	94.2	22.9	99.92	99.99	2.42	>88.0
Potassium, g/100 g[9; 53.9 %]	49.9	13.7	46.9	97.6	34.5	62
Chloride (as NaCl), g/100 g	87.6	12.4	91.4	99.3	44.8	Not indicated

[a]Numbers in parenthesis represent the number (*n*) of samples with concentrations beneath the guaranteed analysis for a specific element. Concentrations considered were those which are not within the lower guaranteed range allowing for one SD (σ). [b]All samples for this mineral are within the guaranteed range

manufacturer or *in situ*), this issue seems to arise more frequently with selenium mineral sources. For example, samples branded as Microgran™ Se and sodium selenite were in some cases inverted i.e. the latter showed low sodium and selenium concentrations (≈5.0-10.0 g/100 g) and viceversa; a reagent grade NaSeO₃ with 99 % purity, when analyzed, should return values for sodium and selenium of at least 13.1 and 45.2 g/100 g, respectively (data not shown). A couple of the recollected samples, considered as manganese (IV) oxide, exhibited only concentrations ranging on the low side of the mg/kg of Mn whereas one might expect for a reagent grade MnO_2 at least 62.6 g/100 g of the metal (data not shown). Yet another example lies within samples labeled as zinc oxide (*n* = 2) in which Zn analysis verified concentrations > 90 g/100 g of this metal which is, chemically speaking, an impossibility in terms of formula, this suggest these samples were not metal oxides but other species entirely. The same occurs in a sample (*n* = 1) of $CoCO_3$ (82.4 g Co/100 g) and a sample (*n* = 1) of KCl (97.58 g K/100 g) (Table 1).

Some samples exhibit mean values significantly (*p* < 0.001) beneath the guaranteed analysis, important examples include KCl for potassium, MgO for magnesium, EDDI for iodine, ferrous and calcium carbonate for iron and carbonate, respectively (Table 1). Congruently, 16.8 % a total of the samples (*n* = 73) were found to be in this

condition. Incidence of some relevant samples was as follow: 83.6 (*n* = 5), 53.8 (*n* = 7), 52.9 (*n* = 9), 47.1 (*n* = 8), 35.0 (*n* = 7) and 28.6 % (*n* = 6) for EDDI, $CoCO_3$, KCl, $ZnSO_4$, MnO and MgO respectively (Table 1). Other minerals do not differ significantly or are between manufacturer assured ranges. Only for minerals Microgran™ Co (*n* = 14) and $MgSO_4$ (*n* = 17) all samples were within the expected concentration range (Table 1). In this regard, mineral deficiencies should be circumvented as these nutrients are relevant to health and their monitoring would consequently could prevent or manage mineral-associated deficiency diseases (Soetan et al. 2010). This is of foremost importance especially in countries when said minerals may be scarce or marginal due to geochemical characteristics of the region. Also, interrelationships and interferences among the mineral elements should be considered when mineral premixes are formulated (Soetan et al. 2010).

Regarding heavy metal concentrations, several mineral samples contravene current legislation in at least one metal. Number of samples in this condition correspond to 0.5 (*n* = 2), 13.8 (*n* = 60), 4.1 (*n* = 18) and 2.5 % (*n* = 11) for As, Hg, Pb and Cd, respectively (Table 2). Overall, 21.1 % (*n* = 92) of the samples exhibited concentrations of heavy metals and fluorine above those stipulated by European guidelines. Values for As, Hg, Pb and Cd ranged from 61.2 mg/kg (MnO) to 1.3 µg/kg (MgO), 2.6×10^3 µg/kg

Table 2 Contaminant concentrations in typical mineral imported supplements employed in feedingstuff manufacture

Assay[a]	Mean	SD	Median	Max	Min	Guideline value or guaranteed analysis
Calcium and phosphorus-containing mineral sources						
Dicalcium phosphate, n = 22						
[b]Arsenic, mg/kg[2; 9.1 %] [4; 18.2 %]	2.88	7.14	0.30	26.68	9.8×10^{-3}	<10
[b]Mercury, µg/kg[8; 36.4 %]	96.18	80.69	71.00	392.50	6.89	<100
Lead, mg/kg	1.89	1.50	1.51	6.23	4.8×10^{-2}	<15
Cadmium, mg/kg	1.94	2.46	0.29	6.73	1.8×10^{-3}	<10
[b]Fluorine ($\times 10^{-2}$), g/100 g[1; 4.5 %]	8.9	4.6	8.2	26.0	2.1	12.0 – 21.0 (<20.0)
Monocalcium phosphate, n = 22						
Arsenic, mg/kg[4; 18.2 %]	3.9×10^{-1}	5.8×10^{-1}	1.1×10^{-1}	2	4.9×10^{-2}	<10
[b]Mercury, µg/kg[12; 54.5 %]	118.85	72.04	110.73	364.00	21.10	<100
[b]Lead, mg/kg[2; 9.1 %] [2; 9.1 %]	38.16	153.47	0.48	706.00	5.4×10^{-2}	<15
Cadmium, µg/kg	2.76	1.82	2.19	6.96	1.6×10^{-3}	<10
Fluorine ($\times 10^{-2}$), g/100 g	7.6	4.4	9.8	14.0	1.0	<0.21 (<0.20)
Calcium carbonate, n = 21						
Arsenic, µg/kg[7; 33.3 %]	383.4	199.4	362.7	818.5	50.9	$<1.5 \times 10^{4}$
Mercury, µg/kg[5; 23.8 %]	73.0	45.6	72.0	185.8	5.6	<300
[b]Lead, mg/kg[2; 9.5 %] [1; 4.8 %]	14.4	60.5	0.2	270.9	1.4×10^{-2}	<20
Cadmium, µg/kg	647.7	559.3	382.9	1.7×10^{3}	1.3	$<1.0 \times 10^{4}$
Iodine-containing mineral sources						
Calcium iodate, n = 7						
Arsenic, µg/kg[1; 14.3 %]	244.9	143.15	196.0	451.4	83.0	$<1.5 \times 10^{4}$
[b]Mercury, µg/kg[1; 14.3 %] [1; 14.3 %]	164.8	103.75	116.4	334.0	64.99	<300
[b]Lead, mg/kg[1; 14.3 %] [1; 14.3 %]	12.3	23.5	0.5	64.4	3.5×10^{-2}	<20
Cadmium, µg/kg[2; 28.6 %]	47.4	34.0	41.9	110.5	10.3	$<2.0 \times 10^{3}$
Microgran™ I 10 % BMP (calcium iodide), n = 10						
Arsenic, mg/kg[3; 30.0 %]	0.69	1.32	0.18	3.92	3.0×10^{-2}	<30
Mercury, µg/kg[1; 10.0 %]	98.5	40.6	115.8	143.68	17.5	<200
Lead, µg/kg[3; 30.0 %]	183.7	98.9	163.0	363.0	55	$<1.0 \times 10^{5}$
Cadmium, µg/kg	146.1	18.9	149.3	168.4	112.9	$<1.0 \times 10^{4}$
Etilendiamine diiodide, n = 7						
Arsenic, µg/kg[2; 28.7 %]	65.0	42.3	64.1	118	13.98	$<3.0 \times 10^{4}$
Mercury, µg/kg[1; 14.3 %]	104.3	31.0	116.1	131.3	45.7	<200
Lead, µg/kg[1; 14.3 %]	170.9	73.1	148.0	301.0	78.0	$<1.0 \times 10^{5}$
Cadmium, µg/kg[2; 28.7 %]	319.8	502.5	35.7	1190	17.5	$<1.0 \times 10^{4}$
Potassium Iodide, n = 7						
Arsenic, µg/kg[2; 28.6 %]	237.9	171.0	156.5	509.6	57.0	$<3.0 \times 10^{4}$
Mercury, µg/kg	117.7	71.9	102.2	245.7	10.3	<200
[b]Lead, mg/kg[2; 28.6 %], [1; 14.3 %]	42.3	83.7	0.25	209.6	5.7×10^{-2}	<100
Cadmium, µg/kg[1; 14.3 %]	150.8	163.8	35.0	382.4	34.9	$<1.0 \times 10^{4}$
Cobalt-containing mineral sources						
Cobalt carbonate, n = 13						
Arsenic, mg/kg[5; 38.5 %]	7.4×10^{-1}	1.6	9.5×10^{-2}	5.10	1.3×10^{-2}	<30
[b]Mercury, µg/kg[1; 7.7 %]	97.3	69.5	97.5	302.0	23.9	<200
Lead, µg/kg[3; 23.1 %]	664.5	582.8	414.0	1920.0	2.3	$<1.0 \times 10^{5}$

Table 2 Contaminant concentrations in typical mineral imported supplements employed in feedingstuff manufacture *(Continued)*

Assay[a]	Mean	SD	Median	Max	Min	Guideline value or guaranteed analysis
Cadmium, μg/kg[(6; 46.1 %)]	34.0	19.6	38.0	55.7	8.3	$<5.0 \times 10^3$
Cobalt sulfate, n = 11						
Arsenic, μg/kg[(3; 27.3 %)]	65.2	61.4	32.8	151.2	18.0	$<3.0 \times 10^4$
Mercury, μg/kg	81.7	29.6	80.0	133.7	28.2	<200
Lead, mg/kg[(2; 18.2 %)]	17.6	37.3	1.26	101.0	7.0×10^{-3}	<100
Cadmium, μg/kg	835.3	880.8	192.3	2.2×10^3	66.1	$<5.0 \times 10^3$
Microgran™ Co 5 % BMP (cobalt carbonate), n = 13						
Arsenic, mg/kg[(8; 61.5 %)]	1.07	1.65	0.36	4.72	3.42×10^{-2}	30
[b]Mercury, mg/kg[(1; 7.7 %), [2; 15.4 %]]	95.0	58.0	83.1	220.7	37.0	<200
Lead, mg/kg[(4; 30.8 %)]	16.2	48.2	1.4×10^{-1}	160.7	1.8×10^{-3}	<100
Cadmium, μg/kg	84.3	78.2	34.1	239.2	0.95	$<5.0 \times 10^3$
Copper-containing mineral sources						
Copper oxide, n = 19						
Arsenic, mg/kg[(1; 5.3 %)]	2.38	2.19	2.25	8.38	1.29×10^{-1}	100
[b]Mercury, mg/kg[(1; 5.3 %), [3; 15.8 %]]	128.4	88.8	99.1	384.5	48.3	<200
[b]Lead, mg/kg[[12; 63.1 %]]	96.0	113.2	60.9	455.0	3.80×10^{-1}	<30
[b]Cadmium, mg/kg[(1), [3; 15.8 %]]	3.09	2.31	1.85	7.41	1.09×10^{-1}	<5
Copper sulfate, n = 18						
Arsenic, μg/kg[(1; 5.5 %)]	500.6	537.4	366.3	2.29×10^3	18.4	5.0×10^4
Mercury, μg/kg	69.4	35.1	69.4	118.3	9.04	<200
Lead, mg/kg[(2; 11.1 %)]	9.9	10.2	8.8	39.8	4.4×10^{-2}	<100
Cadmium, μg/kg[(4; 22.2 %)]	544.0	1.2×10^3	150.8	4.1×10^3	4.41	$<5.0 \times 10^3$
Sodium containing mineral sources						
Sodium chloride, n = 23						
Arsenic, mg/kg[(16; 69.6 %)]	101.1	61.1	97.5	231.0	15.7	30
Mercury, μg/kg[(4; 17.4 %)]	80.6	43.4	85.0	157.0	1.2	<200
[b]Lead, mg/kg[[1; 4.5 %], (15; 65.2 %)]	26.2	68.8	1.7×10^{-1}	208.4	4.6×10^{-2}	<100
Cadmium, mg/kg[(5; 21.3 %)]	1.21	3.12	3.4×10^{-2}	9.46	4.2×10^{-3}	<5
Iron-containing mineral sources						
Ferrous carbonate, n = 21						
Arsenic, mg/kg	6.50	5.07	5.55	19.4	1.15	30
[b]Mercury, μg/kg[[3; 14.3 %]]	149.0	142.9	110.2	594.0	8.3	<200
Lead, mg/kg	14.51	17.35	7.44	72.06	1.1×10^{-2}	<200
Cadmium, mg/kg[(2; 9.5 %)]	0.73	1.36	1.0×10^{-1}	4.45	7.7×10^{-3}	<5
Ferric oxide, n = 16						
Arsenic, mg/kg[(1; 6.2 %)]	6.38	5.28	4.89	18.49	1.7×10^{-2}	30
[b]Mercury, μg/kg[[3; 18.7 %]]	155.1	126.1	122.1	603.0	26.2	<200
Lead, mg/kg	8.94	13.57	2.67	49.18	1.14	<100
[b]Cadmium, mg/kg[(4; 25.0 %), [1; 6.2 %]]	0.82	1.95	7.0×10^{-2}	6.31	1.1×10^{-2}	$<5.0 \times 10^3$
Ferrous sulfate, n = 17						
Arsenic, μg/kg[(4; 23.5 %)]	1323.7	3395.1	118.2	12750	3.02	3.0×10^4
Mercury, μg/kg	169.4	319.7	100.2	1439.0	23.8	<200
[b]Lead, mg/kg[(2; 11.8 %), [1; 5.9 %]]	34.8	96.1	8.8	393.00	3.1×10^{-2}	<100
Cadmium, μg/kg[(6; 35.3 %)]	45.8	30.7	45.1	107.1	6.1	$<5.0 \times 10^3$

Table 2 Contaminant concentrations in typical mineral imported supplements employed in feedingstuff manufacture *(Continued)*

Assay[a]	Mean	SD	Median	Max	Min	Guideline value or guaranteed analysis
Magnesium-containing mineral sources						
Magnesium oxide, n = 21						
Arsenic, μg/kg[8; 38.1 %]	944.4	1289.9	559.4	5200.0	1.3	2.0×10^4
[b]Mercury, μg/kg[1; 4.8 %), [2; 9.5 %]	81.1	72.1	68.1	287.0	6.6	<200
Lead, mg/kg[11; 52.4 %), [1; 4.8 %]	39.0	135.8	7.7×10^{-1}	628.1	6.0×10^{-2}	<100
Cadmium, μg/kg[4; 19.0 %]	54.4	79.6	28.0	262.1	7.7	$<5.0 \times 10^3$
Magnesium sulfate, n = 18						
Arsenic, μg/kg[8; 44.4 %]	288.1	297.8	75.1	784.7	7.38	2.0×10^3
Mercury, μg/kg	86.0	34.9	79.1	164.1	5.02	<200
Lead, mg/kg[11; 61.1 %]	5.1	11.5	2.4×10^{-1}	33.2	2.2×10^{-2}	<100
Cadmium, μg/kg[10; 55.5 %]	18.5	13.7	10.4	40.6	3.6	$<5.0 \times 10^3$
Manganese-containing mineral sources						
Manganous oxide, n = 21						
Arsenic, mg/kg	28.6	18.3	29.0	61.2	1.2	100
[b]Mercury, μg/kg [6; 28.6 %]	348.6	543.0	198.1	2.6×10^3	1.23	<200
Lead, mg/kg	28.5	19.6	26.1	79.8	2.7×10^{-2}	<200
[b]Cadmium, mg/kg[1; 4.8 %), [6; 28.6 %]	2.84	2.25	3.47	5.49	2.0×10^{-3}	<5
Manganese sulfate, n = 15						
Arsenic, μg/kg[4; 26.7 %]	418.8	470.2	271.9	1.6×10^3	18.8	2.0×10^4
Mercury, μg/kg	67.2	41.9	69.3	163.4	6.2	<200
Lead, μg/kg[8; 53.3 %]	1.3×10^3	2.1×10^3	265.2	5.0×10^3	6.3	$<1.0 \times 10^5$
Cadmium, μg/kg[2; 13.3 %]	1.1×10^3	1.5×10^3	287.2	4.0×10^3	48.11	$<5.0 \times 10^3$
Sulfur-containing mineral sources						
Elemental sulfur, n = 20						
Arsenic, μg/kg[15; 75.0 %]	76.1	38.0	71.2	145.8	39.4	2.0×10^4
[b]Mercury, μg/kg[6; 30.0 %]	21.4	52.1	1.7×10^{-1}	149.0	2.6	<200
Lead, mg/kg[13; 65.0 %]	1.6×10^{-1}	5.2×10^{-2}	1.6×10^{-1}	2.4×10^{-1}	6.9×10^{-2}	<100
Cadmium, μg/kg[7; 35.0 %]	30.6	15.4	30.3	52.4	4.5	$<5.0 \times 10^3$
Zinc-containing mineral sources						
Zinc oxide, n = 22						
Arsenic, μg/kg[1; 4.5 %]	1.3×10^3	2.3×10^3	365.2	9.8×10^3	12.75	$<1.0 \times 10^5$
[b]Mercury, μg/kg[2; 9.1 %]	133.1	181.0	78.9	882.0	16.1	<200
Lead, mg/kg	55.8	68.2	26.5	252.0	2.5×10^{-2}	<400
[b]Cadmium, μg/kg[1; 4.5 %), [1; 4.5 %]	2.5×10^3	1.7×10^3	2.8×10^3	5.0×10^3	5.03	$<5.0 \times 10^3$
Zinc sulfate, n = 18						
Arsenic, μg/kg[11; 61.1 %]	347.4	667.2	83.9	2.0×10^3	15.79	2.0×10^4
Mercury, μg/kg[1; 5.5 %]	95.6	44.8	81.4	186.6	28.7	<200
Lead, mg/kg[8; 44.4 %]	2.9×10^{-1}	2.3×10^{-1}	2.1×10^{-1}	7.0×10^{-1}	5.9×10^{-2}	<100
[b]Cadmium, μg/kg[1; 5.5 %]	2.7×10^3	2.9×10^3	1.8×10^3	1.2×10^4	1.75	$<5.0 \times 10^3$
Selenium-containing mineral sources						
Microgran™ Se 1 % BMP (sodium selenite), n = 13						
Arsenic, μg/kg [5; 38.5 %]	544.7	721.8	310.8	2.4×10^3	10.2	$<2.0 \times 10^4$
[b]Mercury, μg/kg[4; 30.8 %]	148.4	126.7	118.9	457.5	28.0	<200
Lead, μg/kg[9; 69.2 %]	6.1×10^{-1}	6.0×10^{-1}	4.0×10^{-1}	1.6	2.4×10^{-2}	$<1.0 \times 10^5$

Table 2 Contaminant concentrations in typical mineral imported supplements employed in feedingstuff manufacture *(Continued)*

Assay[a]	Mean	SD	Median	Max	Min	Guideline value or guaranteed analysis
Cadmium, µg/kg	133.5	82.8	162.2	230.8	17.1	$<5.0 \times 10^3$
Sodium selenite 1 %, n = 11						
Arsenic, µg/kg[5; 45.5 %]	1.4×10^3	9.6×10^2	1.4×10^3	2.8×10^3	172.9	2.0×10^4
[b]Mercury, µg/kg[3; 27.3 %]	165.8	87.5	146.1	330.7	33.44	<200
Lead, µg/kg[1; 9.1 %]	32.5	50.6	7.2×10^2	1.1×10^3	13.3	1.0×10^5
Cadmium, µg/kg	625.2	344.9	725.7	1.1×10^3	13.3	$<5.0 \times 10^3$
Sodium selenite, n = 8						
Arsenic, µg/kg[2; 25.0 %]	3.1×10^3	2.3×10^3	2.9×10^3	7.2×10^3	6.0×10^2	2.0×10^4
[b]Mercury, µg/kg[4; 50.0 %]	264.0	146.4	241.5	462.0	93.9	<200
Lead, mg/kg[4; 50.0 %]	3.79	3.36	3.45	8.16	1.5×10^{-1}	<100
Cadmium, µg/kg[5; 62.5 %]	34.8	10.1	34.8	44.9	24.7	$<5.0 \times 10^3$
Potassium-containing mineral sources						
Potassium Chloride, n = 17						
Arsenic, µg/kg[10; 58.8 %]	188.1	210.9	124.3	672.2	9.6	2.0×10^4
Mercury, µg/kg[1; 5.9 %]	85.3	46.4	81.1	181.3	3.9	<200
Lead, µg/kg[6; 35.6 %]	69.3	110.1	4.5×10^{-1}	323.1	5.5×10^{-2}	1.0×10^5
Cadmium, µg/kg[8; 47.0 %]	120.9	187.1	50.1	492.7	3.4×10^{-1}	$<5.0 \times 10^3$

[a]Numbers in parenthesis represent the number (*n*) of samples that are not detected using our current method of analysis. [b]Represent minerals for which at least one sample's concentration is above the European guidelines and square brackets represent the actual number of samples in this situation

(MnO) to 1.2 µg/kg (NaCl), 706.2 mg/kg [Ca(H$_2$PO$_4$)$_2$] to 5.5×10^{-2} mg/kg (KCl) and 9.46 mg/kg (NaCl) to 3.4×10^{-1} µg/kg (KCl), respectively (Table 2). Other especially elevated Hg and Pb concentrations (over the 200 µg/kg and 100 mg/kg respective permitted levels) found during our survey are 1439 (FeSO$_4$), 882 (ZnO) and 603 (Fe$_2$O$_3$) µg/kg and 628.1 (MgO) and 455 (CuO) mg/kg, respectively (Table 2).

In decreasing order of heavy metal overall concentrations the following minerals showed the less contamination: MnSO$_4$ < EDDI < KCl (Table 2). Microgran™ Se, Co and I samples also showed relatively lower concentrations of heavy metals. This may be expected as this ingredient's presentation mineral input is from 1 to 10 g/100 g, maximum. The fact that the one iodine source from organic synthesis has a lower concentration of heavy metals, sustains our hypothesis that higher contents come from inorganic synthesis raw materials. In the light of this findings other organic sources could be examined, especially since iodine organic salts have demonstrated potential as additives. Actually, iodine consumption through feed (pet and cattle mostly) is primarily as the compound EDDI (Lyday 2005).

Of the heavy metals assayed As showed, in general, the lowest values. These results are especially reassuring considering that all livestock species are susceptible to toxic effects of inorganic arsenic and some feeds may even contain organoarsenical species (e.g. roxarsone) as growth promoters to improve feed efficiency (Chapman and Johnson 2002), occasionally in combination with ionophores. Strikingly, As has been suggested to possess some essential or beneficial functions at ultra-trace concentrations (Uthus 2003).

There is considerably less incidence of relatively elevated As concentrations, however As regulation is in some cases 100 fold more permissive with respect to Hg. In contrast, the higher number of incidents of irregular concentrations for Hg is evident. This is expected as Hg is the most abundant naturally-occurring heavy metal and is emitted primarily due industrial sources and mining ore deposits (Goyer 1996). As most information on mercury residues in feedstuffs, data presented here is given as total mercury concentrations. In this regard, although inorganic mercury toxicity profile due to accumulation include kidney damage, methylmercury (CH$_3$-Hg) is the form considered of greatest toxicological concern which very well be non-existent, considering the nature of the samples tested.

Moreover, up to 30.0 % (Table 2) of samples of sulfur assayed (*n* = 20) showed residues of Hg > 200 µg/kg. This result may be explained by sulfur chemisorption capability of Hg (Feng et al. 2006). In turn, the same samples showed relatively low concentrations of the other heavy metals despite its tendency to associate with them.

Overall, calcium phosphates and metal oxides showed a significantly higher ($p < 0.001$) levels of arsenic and lead relative to other mineral sources (Table 2). This fact could be explained due to the fact that minerals such as Zn, Mn and Fe oxides have such redox potentials that can oxidize As and will thus alter the extent of As retention. In fact, Mn^{3+}/Mn^{5+} oxides are strong oxidants that can oxidize and sequester many trace metals found in nature (García-Sánchez et al. 1999). The capacity of these two minerals in terms of As adsorption will be determined by their adsorption isotherms and the origin of the mineral (or its parent compounds during synthesis) and the predominant species of arsenic found. Ferrous carbonate and manganese sulfate seem to have the same ability. Other researchers already have established a stronger association between lead and metal oxides with respect to other ions and an adsorption of lead by the manganous oxides was up to 40 times greater than that by the iron oxides (McKenzie 1980). This result seems to be the case for our analysis as well; levels of As and Pb are significantly ($p < 0.05$) lower for ZnO (mean values of 1.3×10^3 µg/kg and 55.8 mg/kg, respectively) and higher for FeO (mean values of 6.38×10^3 µg/kg and 155.1 mg/kg, respectively) (Table 2). The latter mineral is of special importance as is known, in some cases, to be used as a pigment (Potter 2000) hence its input in compound feed may be higher relative to other minerals. Heavy metals have been associated with a high sorption in metal oxides even in environmental samples such as sediments (Brown and Parks 2001).

Congruently, dicalcium- and monocalcium- phosphate samples showed the highest frequency in contaminated samples $n = 10/22$ and $n = 14/22$, respectively (Table 2). In this specific case, the presence of arsenic could indicate a certain degree of substitution between arsenate and phosphate ions in the lattice of the calcium salt (Tawfik and Viola 2011) or co-precipitation of the arsenate oxyanion in the presence of calcium phosphate (Clara and Maglhães 2002; Sahai et al. 2007; Henke 2009). In this case, Pb and Cd concentrations in monocalcium phosphate are significantly higher ($p < 0.001$ and $p < 0.05$, respectively) than in dicalcium phosphate. The converse is true for As ($p < 0.001$). Hg concentrations showed no significant differences between both minerals ($p < 0.05$).

On the other hand, despite the relatively lower concentrations of Cd found in the samples, this metal has been reported (Chaney and Ryan 1994; Chaney et al. 1999; Li et al. 2005) to have the greatest potential for transmission through the food chain at levels that present risk to the final consumer.

One key aspect that follows from the regulatory and food safety standpoint, is the amount of heavy metals that is ingested as a result from the mixture of several of these mineral feeds and ingredients. For example, if a compound feed for swine is manufactured with dicalcium phosphate and sodium chloride as metal sources exhibiting the maximum concentration of, say, mercury detected in them and contains ca. 0.8 g/100 g Ca and 0.4 g/ 100 g salt (a common formulation for swine nutrition and are components with relatively high concentrations in feed) assuming this two as the only sources of Hg then this feed will have a total of the metal of 0.38 µg/kg. If a pig of 22 weeks of age was fed with 4 000 g of such feed daily, then it would ingest 15 µg mercury every 24 h.

Even though the amount may seem small, it must be taken into account that health effects from this substance exposure are chronic events, taking time and repeated exposure for the contaminants to bioaccumulate up to toxic levels. Hence, animals with longer life spans may exhibit higher concentrations of heavy metals in their tissues. As a result of this bioaccumulation, the consumption of meat from older animals could represent an increased risk for ingestion to a final consumer. This may suggest animals to be vectors in heavy metal transmission along the food chain (Pagán-Rodríguez et al. 2007).

However, considering that several of these minerals are added to a compound feed, in different proportions, not only the additive character of the concentrations of these heavy metals should be addressed, but also the possibility of dilution by other individual components with lower concentrations of said metals (e.g. maize mill) which results in a relatively low heavy metal containing feed (Table 3). We did not find any of the samples assayed ($n = 50$) [bovine ($n = 10$), fish ($n = 10$), poultry ($n = 10$), shrimp ($n = 10$) and swine ($n = 10$) feeds from the most important production facilities across the country] to surpass current legislation; in fact maximum levels of contaminants found were of 156.7 µg/kg (Table 3).

Table 3 Heavy metal concentrations found in Costarican animal feed samples

Metal/Feed type	Concentration range, µg/kg[a,b]					
	Bovine	Fish	Poultry	Shrimp	Swine	Guideline
Arsenic	Not detected	19.3 – 25.4	48.5 – 90.4	37.8 – 53.5	7.7 – 30.5	2000
Mercury	60.2 – 186.8	50.0 – 55.9	35.9 – 38.6	39.3 – 41.8	38.5 – 71.4	100
Lead	Not detected	125.7 – 127.0	Not detected	Not detected	132.9 – 151.7	5000
Cadmium	113.1 – 156.7	39.9 – 40.3	35.3 – 107.9	19.3 – 35.0	59.5 – 61.2	1000/500

[a]Values obtained from $n = 10$. [b]For all cases SD < 5 %; three replicates per sample

These metals are of special importance as they are usually documented as substances with strong toxigenic and carcinogenic capacities (van Paemel et al. 2010). For example, both As^{3+} and As^{5+} are classified as group A human carcinogens (US EPA 1998; Sapkota 2007). This is of foremost importance in mineral feed since most vitamins and minerals are "generally recognized as safe" according with food additive regulation (US FDA 2014). However, some protection policies warnings to the feed industry have been issued against the use of mineral sources that are by-products or co-products of industrial metal production (US FDA 2003).

Considering the data provided herein, a programme that strictly monitor the quality on mineral ingredients and heavy metal concentrations should be implemented in order to maintain toxic metal residues within acceptable levels and avoid contaminated feed ingredients entering the food chain. This is especially relevant for cadmium and lead (considering their prevalence in the environment) and mercury contemplating the relatively elevated values found during our survey. Trends in concentration of contaminants in feed and feed ingredients should also be observed and from the animal and human health standpoint, interactions between heavy metals and essential nutrients (D'Souza et al. 2003) should be also considered.

Noteworthy, thanks to our work and the data compiled here, several Costa Rican feed manufacturers had taken steps on improving their manufacturing practices and had avoided all together the use of several raw materials with recurrent irregular mineral or heavy metal concentrations. Finally, we recommend further research include speciation as toxicity of the heavy metal involved is closely related to its oxidation state.

Conclusions

Occasionally, mineral ingredient samples surpass, in some cases with elevated concentrations, permitted levels of undesirable substances in this case specifically, toxic metals. As the relative frequency is relatively high, a strict monitoring programme of both main composition and toxic metals must be sustained regularly in order to guarantee both the quality and safety of the ingredients used in animal feeds. Finally, mineral values of these raw materials on some instances were lower than declared possibly due to errors in manufacturing, local shipping, handling or in-plant repackaging.

Abbreviations

AAFCO: Association of american feed control officials; AAS: Atomic absorption spectrometry; ACS: American chemical society; AOAC: association of official analytical chemists; EC: European community; EDDI: ethylenediamine dihydroiodide; EPA: Environmental protection agency; FAAS: Flame atomic absorption spectrometry; FDA: Food and drug administration; FI: Flow injection; GFAAS: Graphite furnace atomic absorption spectrometry; OMA: Official methods of analysis; MH: Metal hydride; TISAB: Total ionic strength adjustment buffer; SD: Standard deviation; USD: United States dollars; UV: Ultraviolet.

Competing interests
The authors declare that they have no competing interests.

Author's contributions
LM, SP and FG designed the experiment; LM recollected the samples and devised the sampling design; FG and SP designed the analysis approach and performed measurements; FG, SP and LM analyzed the data; FG wrote the draft of the manuscript. All authors read and approved the final manuscript.

Acknowledgements
The authors would like to thank the plants and their respective representatives who were willing to participate in this research proposal. Marian Flores and Andrea Porras are acknowledged for their technical assistance. Vicerrectoría de Investigación supported this initiative by means of project B2059.

Author details
[1]Centro de Investigación en Nutrición Animal (CINA), Universidad de Costa Rica, Ciudad Universitaria Rodrigo Facio, San José 11501-2060, Costa Rica. [2]Escuela de Zootecnia, Universidad de Costa Rica, Ciudad Universitaria Rodrigo Facio, San José 11501-2060, Costa Rica.

References
Association of American Feed Control Officials (2014) Official Publication, Association of American Feed Control. Officials, Oxfordshire
Blanco-Penedo I, Shore RF, Miranda M, Benedito JL, López-Alonso M (2009) Factors affecting trace elements status in calves in NW Spain. Livest Sci 123:198–208
Brown GE Jr, Parks GA (2001) Sorption of trace elements on mineral surfaces: Modern perspectives from spectroscopic studies, and comments on sorption in the marine environment. Int Geol Rev 43:963–1073
Chaney RL, Ryan JA (1994) Risk Based Standards for Arsenic, Lead and Cadmium in Urban Soils. Dechema, Frankfurt
Chaney RL, Ryan JA, Brown SL (1999) Environmentally Acceptable Endpoints for Soil Metals. In: Loehr RC, Anderson WC, Smith BP (eds) Environmental availability in soils: Chlorinated Organics, Explosives, Metals, 1st edn. Am Acad Environ, Annapolis
Chapman HD, Johnson ZB (2002) Use of antibiotics and roxarsone in broiler chickens in the USA: analysis for the years 1995 to 2000. Poult Sci 81:356–364
Clara M, Magalhães F (2002) Arsenic. An environmental problem limited by solubility. Pure Appl Chem 74:1843–1850
Demirezen D, Uruç K (2006) Comparative study of trace elements in certain fish, meat and meat products. Meat Sci 74:255–260
dos Santos AP, Diehl A, Levin Y (2010) Surface tensions, surface potentials and the Hofmeister series of electrolyte solutions. Langmuir 26:10778–10783
D'Souza HS, Menezes G, Venkatesh T (2003) Role of essential trace minerals on the absorption of heavy metals with special reference to lead. Indian J Clin Biochem 18:154–160
Elzea J, Trivedi NC, Barker JM, Krukowski ST (eds) (2006) Industrial Minerals and rocks Commodities, market and uses, 7th edn. Society for Mining, Metallurgy, and Exploration, Colorado
[EPA] U.S. Environmental Protection Agency (1998) Integrated Risk Information System. Arsenic, Inorganic (CASRN 7440-38-2). http://www.epa.gov/iris/subst/0278.htm. Accessed 12 December 2014
European Parliament and of the Council (2000) White Paper on Ford Safety COM/99/0719 final. http://eur-lex.europa.eu/legal-content/EN/TXT/PDF/?uri=CELEX:51999DC0719&from=ES. Accessed 06 April 2015
European Parliament and of the Council (2002) Directive 2002/32/EC On undesirable substances in animal feed. http://europa.eu/legislation_summaries/food_safety/animal_nutrition/l12069_en.htm. Accessed 12 December 2014
European Parliament and of the Council (2007) Directive 2007/333/EC On undesirable substances in animal feed. http://eur-lex.europa.eu/LexUriServ/LexUriServ.do?uri=OJ:L:2007:088:0029:0038:EN:PDF. Accessed 05 April 2015
European Commission Joint Research Centre (2012) Official methods for the determination of heavy metals in feed and food. https://ec.europa.eu/jrc/sites/default/files/Official%20methods%20for%20the%20determination%20of%20heavy%20metals%20in%20feed%20and%20food_v4.pdf. Accessed 05 April 2015

[FDA] U.S. Food and Drug Administration (2003) CVM Update, FDA Information for Manufacturers of Animal Feed Mineral Mixes. Center for Veterinary Medicine, March 12, 2003

[FDA] U.S. Food and Drug Administration (2014) Code of Federal Regulations. http://www.accessdata.fda.gov/scripts/cdrh/cfdocs/cfcfr/CFRSearch.cfm?fr=573.870. Accessed 20 December 2014

Feng W, Borguet E, Vidic RD (2006) Sulfurization of a carbon surface for vapor phase mercury removal – II: Sulfur forms and mercury uptake. Carbon 44:2998–3004

Fernandes JI, Lima FR, Mendonça CX Jr, Mabe I, Albuquerque R, Leal PM (1999) Relative bioavailability of phosphorus in feed and agricultural phosphates for poultry. Poult Sci 12:1729–1736

García-Sánchez A, Alastuey A, Querol X (1999) Heavy metal adsorption by different minerals: Application to the remediation of polluted soils. Sci Total Environ 242:179–188

Goyer RA (1996) Toxic effects of metals: mercury. In: Casarett and Doull's Toxicology: The Basic Science of Poisons, 5th edn. McGraw-Hill, New York

Hasty RA (1973) Reduction of iodate by hydrazine: Application of the iodide ion selective electrode to the uncatalyzed reaction. Mickrochimika acta 61:925–933

Henke KR (2009) Arsenic: Environmental Chemistry, Health Threats and Waste Treatment. In: Henke KR (ed) Arsenic in natural environments. John Wiley & Sons, Inc, Sussex

Hoffmann J, Hoffmann K, Skut J, Huculak-Mączka M (2011) Modification of manufacturing process of feed phosphates. CHEMIK 65:184–191

Järup L (2003) Hazards of heavy metal contamination. Br Med Bull 68:167–182

Leeman WR, van der Berg KJ, Houben GF (2007) Transfer of chemicals from feed to animal products: The use of transfer factors in risk assessment. Food Addit Contam 24:1–13

Li Y, McCrory DF, Powell JM, Saam H, Jackson-Smith D (2005) A Survey of selected Heavy metal concentrations in Winsconsin Dairy Feeds. J Dairy Sci 88:2911–2922

López-Alonso M, Benedito JL, Miranda M, Castillo C, Hernández J, Shore RF (2002) Contribution of cattle products to dietary intake of trace and toxic elements in Galicia, Spain. Food Addit Contam 19:533–541

Lyday Phyllis A (2005) Iodine and Iodine Compounds. In: Ullmann's Encyclopedia of Industrial Chemistry. Wiley-VCH, Weinheim

Makridis C, Svarnas C, Rigas N, Gougoulias N, Roka L, Leontopoulos S (2012) Transfer of Heavy Metal Contaminants from Animal Feed to Animal Products. J Agr Sci Tech 2:149–154

Mamtani R, Stern P, Dawood I, Cheema S (2011) Metals and Disease: A Global Primary Health Care Perspective. J Toxicol 2011:1–11

Mckenzie RM (1980) The adsorption of lead and other heavy metals on oxides of manganese and iron. Aust J Soil Res 18:61–73

Nicholson FA, Smith SR, Alloway BJ, Carlton-Smith C, Chambers BJ (2003) An inventory of heavy metals inputs to agricultural soils in England and Wales. Sci Total Environ 311:205–219

Official Journal of the European Communities (2012) Commission Regulation (EU) No 744/2012. Off J Eur Union L 140:1–10

Pagán-Rodríguez D, O'Keefe M, Deyrup C, Zervos P, Walker H, Thaler A (2007) Cadmium and lead residue control in a hazard analysis and critical control point (HACCP) environment. J Agric Food Chem 55:1638–1642

Petersen GI, Pedersen C, Lindemann MD, Stein HH (2011) Relative bioavailability of phosphorus in inorganic phosphorus sources fed to growing pigs. J Anim Sci 2:460–466

Plumridge TH, Steele G, Waigh RD (2000) Geometry-based simulation of the hydration of small molecules. Phys Chem Comm 3:36–41

Poirier LA, Littlefield NA (1996) Metal interactions in chemical carcinogenesis. In: Chang LW, Magos L, Suzuki T (eds) Toxicology of Metals, 1st edn. CRC Lewis Publishers, Boca Raton

Potter, MJ (2000) Iron oxide pigments. http://minerals.usgs.gov/minerals/pubs/commodity/iron_oxide/750400.pdf. Accessed 23 November 2014

Promotora del Comercio Exterior de Costa Rica, Portal estadístico de comercio exterior. http://servicios.procomer.go.cr/estadisticas/inicio.aspx. Accessed 23 December 2014

Sahai N, Lee YJ, Xu H, Ciardelli M, Gaillard JF (2007) Role of Fe(II) and phosphate in arsenic uptake by coprecipitation. Geochim Cosmochim Acta 71:3193–3210

Sapkota AR, Lefferts LY, McKenzie S, Walker P (2007) What Do We Feed to Food-Production Animals? A Review of Animal Feed Ingredients and Their Potential Impacts on Human Health. Environ Health Perspect 115:663–670

Sharma RJ, Agrawal M (2005) Biological effects of heavy metals: An overview. J Exp Bot 26:301–313

Singh R, Gautam N, Mishra A, Gupta R (2011) Heavy metals and living systems: An overview. Indian J Pharmacol 43:246–53

Soetan KO, Olaiya CO, Oyewole OE (2010) The importance of mineral elements for humans, domestic animals and plants: A review. Afr J Food Sci 4:220–222

Tawfik DS, Viola RE (2011) Arsenate Replacing Phosphate: Alternative Life Chemistries and Ion Promiscuity. Biochemistry 50:1128–1134

Uthus EO (2003) Arsenic essentiality: A role affecting methionine metabolism. J Trace Elem Exp Med 16:345–355

Van der Weijden WJ, Hees E, Bastein T, Udo de Haes HA (2013) The geopolitics of raw materials for agriculture and food production. Part B: Analysis. Platform Agriculture, Innovation and Society, Culemborg. http://groupedebruges.eu/sites/default/files/publications/downloads/the_geopolitics_of_raw_materials_for_agriculture_and_food_production_-_platform_agriculture_innovation_society_2014_0.pdf. Accessed 06 April 2015

Van Paemel M, Dierick N, Janssens G, Fievez V, de Smet S (2010) Technical report submitted to EFSA: Selected trace and ultratrace elements: Biological role, content in feed and requirements in animal nutrition – Elements for risk assessment. http://www.efsa.europa.eu/en/scdocs/doc/68e.pdf. Accessed 06 April 2015

Viljoen J (2001) Quality of feed phosphate supplements for animal nutrition. SA Anim Sci 2:13–19

Wei ZF, Zhang YH, Zhao LJ, Liu JH, Li XH (2005) Observations of the first hydration layer of isolated cations and anions through the FTIR-ATR difference spectra. J Phys Chem A 109:1334–1337

Permissions

The contributors of this book come from diverse backgrounds, making this book a truly international effort. This book will bring forth new frontiers with its revolutionizing research information and detailed analysis of the nascent developments around the world.

We would like to thank all the contributing authors for lending their expertise to make the book truly unique. They have played a crucial role in the development of this book. Without their invaluable contributions this book wouldn't have been possible. They have made vital efforts to compile up to date information on the varied aspects of this subject to make this book a valuable addition to the collection of many professionals and students.

This book was conceptualized with the vision of imparting up-to-date information and advanced data in this field. To ensure the same, a matchless editorial board was set up. Every individual on the board went through rigorous rounds of assessment to prove their worth. After which they invested a large part of their time researching and compiling the most relevant data for our readers.

The editorial board has been involved in producing this book since its inception. They have spent rigorous hours researching and exploring the diverse topics which have resulted in the successful publishing of this book. They have passed on their knowledge of decades through this book. To expedite this challenging task, the publisher supported the team at every step. A small team of assistant editors was also appointed to further simplify the editing procedure and attain best results for the readers.

Apart from the editorial board, the designing team has also invested a significant amount of their time in understanding the subject and creating the most relevant covers. They scrutinized every image to scout for the most suitable representation of the subject and create an appropriate cover for the book.

The publishing team has been an ardent support to the editorial, designing and production team. Their endless efforts to recruit the best for this project, has resulted in the accomplishment of this book. They are a veteran in the field of academics and their pool of knowledge is as vast as their experience in printing. Their expertise and guidance has proved useful at every step. Their uncompromising quality standards have made this book an exceptional effort. Their encouragement from time to time has been an inspiration for everyone.

The publisher and the editorial board hope that this book will prove to be a valuable piece of knowledge for researchers, students, practitioners and scholars across the globe.

List of Contributors

Faysal Elahi Khan
Department of food Engineering & Tea Technology, Shahjalal University of Science and Technology, Sylhet, Bangladesh

Yeasmin Nahar Jolly
Chemistry Division, Atomic Energy Centre, Dhaka, Bangladesh

GM Rabiul Islam
Department of food Engineering & Tea Technology, Shahjalal University of Science and Technology, Sylhet, Bangladesh

Shirin Akhter
Chemistry Division, Atomic Energy Centre, Dhaka, Bangladesh

Jamiul Kabir
Chemistry Division, Atomic Energy Centre, Dhaka, Bangladesh

Tizeta Bekele
College of Veterinary Medicine and Agriculture, Addis Ababa University, P.O. Box 34, Bishftu, Ethiopia

Girma Zewde
College of Veterinary Medicine and Agriculture, Addis Ababa University, P.O. Box 34, Bishftu, Ethiopia

Genene Tefera
Institute of Biodiversity Conservation, Microbial Genetic Resources Centre, P.O. Box 30726, Addis Ababa, Ethiopia

Aklilu Feleke
College of Veterinary Medicine and Agriculture, Addis Ababa University, P.O. Box 34, Bishftu, Ethiopia

Kaleab Zerom
College of Veterinary Medicine, Samara University, P.O. Box 132, Samara, Ethiopia

Ali Abid Abojassim
Department of Physics, Kufa University, Faculty of Science, Kufa, Iraq

Husain Hamad Al-Gazaly
Department of Physics, Kufa University, Faculty of Science, Kufa, Iraq

Suha Hade Kadhim
Department of Physics, Kufa University, Faculty of Science, Kufa, Iraq

Elizabeth A Jara
Department of Food Science and Technology, University of California, Davis, CA, USA

Carl K Winter
Department of Food Science and Technology, University of California, Davis, CA, USA

Sabrina Bartz
Laboratório de Microbiologia e Controle de Alimentos, Instituto de Ciência e Tecnologia de Alimentos, Universidade Federal do Rio Grande do Sul (ICTA/UFRGS), Av. Bento Gonçalves, 9500, prédio 43212, Campus do Vale, Agronomia, Cep. 91501-970 Porto Alegre/RS, Brazil

Claudia Titze Hessel
Laboratório de Microbiologia e Controle de Alimentos, Instituto de Ciência e Tecnologia de Alimentos, Universidade Federal do Rio Grande do Sul (ICTA/UFRGS), Av. Bento Gonçalves, 9500, prédio 43212, Campus do Vale, Agronomia, Cep. 91501-970 Porto Alegre/RS, Brazil

Rochele de Quadros Rodrigues
Laboratório de Microbiologia e Controle de Alimentos, Instituto de Ciência e Tecnologia de Alimentos, Universidade Federal do Rio Grande do Sul (ICTA/UFRGS), Av. Bento Gonçalves, 9500, prédio 43212, Campus do Vale, Agronomia, Cep. 91501-970 Porto Alegre/RS, Brazil

Anelise Possamai
Laboratório de Microbiologia e Controle de Alimentos, Instituto de Ciência e Tecnologia de Alimentos, Universidade Federal do Rio Grande do Sul (ICTA/UFRGS), Av. Bento Gonçalves, 9500, prédio 43212, Campus do Vale, Agronomia, Cep. 91501-970 Porto Alegre/RS, Brazil

Fabiana Oliveira Perini
Laboratório de Microbiologia e Controle de Alimentos, Instituto de Ciência e Tecnologia de Alimentos, Universidade Federal do Rio Grande do Sul (ICTA/UFRGS), Av. Bento Gonçalves, 9500, prédio 43212, Campus do Vale, Agronomia, Cep. 91501-970 Porto Alegre/RS, Brazil

Liesbeth Jacxsens
Department of Food Safety and Food Quality, Laboratory of Food Preservation and Food Microbiology, Faculty of Bioscience Engineering, Ghent University, Coupure Links, 653, 9000 Ghent, Belgium

Mieke Uyttendaele
Department of Food Safety and Food Quality, Laboratory of Food Preservation and Food Microbiology, Faculty of Bioscience Engineering, Ghent University, Coupure Links, 653, 9000 Ghent, Belgium

Renar João Bender
Laboratório de Pós-Colheita, Faculdade de Agronomia, Universidade Federal do Rio Grande do Sul, Av Bento Gonçalves, 7712. 91540-000 Porto, Alegre/RS, Brazil

Eduardo César Tondo
Laboratório de Microbiologia e Controle de Alimentos, Instituto de Ciência e Tecnologia de Alimentos, Universidade Federal do Rio Grande do Sul (ICTA/UFRGS), Av. Bento Gonçalves, 9500, prédio 43212, Campus do Vale, Agronomia, Cep. 91501-970 Porto Alegre/RS, Brazil

Lawrence I Ezemonye
Ecotoxicology and Environmental Forensics Laboratory, University of Benin, Benin City, Edo State, Nigeria

Ozekeke S Ogbeide
Ecotoxicology and Environmental Forensics Laboratory, University of Benin, Benin City, Edo State, Nigeria

Isioma Tongo
Ecotoxicology and Environmental Forensics Laboratory, University of Benin, Benin City, Edo State, Nigeria

Alex A Enuneku
Ecotoxicology and Environmental Forensics Laboratory, University of Benin, Benin City, Edo State, Nigeria

Emmanuel Ogbomida
National Centre for energy and environment (NCEE), University of Benin, Benin City, Edo State, Nigeria

Shima Shayanfar
National Center for Electron Beam Research, Department of Nutrition and Food Science, Texas A&M University, College Station, Room 418B, Kleberg Center, MS 2472, 77843-2472 Texas, USA

Christina Harzman
BIOTECON Diagnostics GmbH, Hermannswerder 17, 14473 Potsdam, Germany

Suresh D Pillai
National Center for Electron Beam Research, Department of Nutrition and Food Science, Texas A&M University, College Station, Room 418B, Kleberg Center, MS 2472, 77843-2472 Texas, USA

Channa Jayasumana
Faculty of Medicine& Allied Sciences, Rajarata University of Sri Lanka, Saliyapura 50008, Sri Lanka

Priyani Paranagama
Faculty of Science, University of Kelaniya, Colombo 11600, Sri Lanka

Saranga Fonseka
Faculty of Science, University of Kelaniya, Colombo 11600, Sri Lanka

Mala Amarasinghe
Faculty of Science, University of Kelaniya, Colombo 11600, Sri Lanka

Sarath Gunatilake
Department of Health Science, California State University Long Beach, Long Beach, CA 90840, USA

Sisira Siribaddana
Faculty of Medicine & Allied Sciences, Rajarata University of Sri Lanka, Saliyapura 50008, Sri Lanka

Márcia Regina Denadai
Departamento de Biologia Animal, Instituto de Biologia, Universidade Estadual de Campinas, Campinas, SP 13083-862, Brazil
Instituto Oceanográfico, Universidade de São Paulo, São Paulo, SP 05508-900, Brazil

Daniela Franco Carvalho Jacobucci
Instituto de Ciências Biomédicas, Instituto de Biologia, Universidade Federal de Uberlândia, Uberlândia, MG 38400-902, Brazil

Isabella Fontana
Departamento de Medicina Veterinária Preventiva e Saúde Animal, Faculdade de Medicina Veterinária e Zootecnia, Universidade de São Paulo, São Paulo, SP 05508-270, Brazil

Satie Taniguchi
Instituto Oceanográfico, Universidade de São Paulo, São Paulo, SP 05508-900, Brazil

Alexander Turra
Instituto Oceanográfico, Universidade de São Paulo, São Paulo, SP 05508-900, Brazil

Mahbub Murshed Khan
Institute of Nutrition and Food Science, University of Dhaka, Dhaka, Bangladesh

Md Tazul Islam
Department of Food Technology and Nutrition Science, Noakhali Science and Technology University, Noakhali, Bangladesh

Mohammed Mehadi Hassan Chowdhury
Department of Microbiology, Noakhali Science and Technology University, Noakhali, Bangladesh

Sharmin Rumi Alim
Institute of Nutrition and Food Science, University of Dhaka, Dhaka, Bangladesh

Becky Price
Consultant for GeneWatch UK, 60 Lightwood Road, Buxton SK17 7BB, UK

Janet Cotter
Greenpeace Resaerch Laboratories, Innovation Centre Phase 2, University ofExeter, Exeter EX4 4RN, UK

Hany M Yehia
Food Science and Nutrition Department, College of Food and Agricultural Sciences, King Saud University, Riyadh, Saudi Arabia
Food Science and Nutrition Department, Faculty of Home Economics, Helwan University, Cairo, Egypt

Mosffer M AL-Dagal
Food Science and Nutrition Department, College of Food and Agricultural Sciences, King Saud University, Riyadh, Saudi Arabia

Marijke Verhegghe
Institute for Agricultural and Fisheries Research (ILVO), Technology and Food Science Unit, Food safety research group, Brusselsesteenweg 370, Melle 9090, Belgium
Department of Pathology, Bacteriology and Avian Diseases, Ghent University, Faculty of Veterinary Medicine, Salisburylaan 133, Merelbeke 9820, Belgium

Lieve Herman
Institute for Agricultural and Fisheries Research (ILVO), Technology and Food Science Unit, Food safety research group, Brusselsesteenweg 370, Melle 9090, Belgium

Freddy Haesebrouck
Department of Pathology, Bacteriology and Avian Diseases, Ghent University, Faculty of Veterinary Medicine, Salisburylaan 133, Merelbeke 9820, Belgium

Patrick Butaye
Department of Pathology, Bacteriology and Avian Diseases, Ghent University, Faculty of Veterinary Medicine, Salisburylaan 133, Merelbeke 9820, Belgium
Department of Bacteriology and Immunology, Veterinary and Agrochemical Research Centre (VAR), Groeselenberg 99, Brussels 1180, Belgium

Marc Heyndrickx
Institute for Agricultural and Fisheries Research (ILVO), Technology and Food Science Unit, Food safety research group, Brusselsesteenweg 370, Melle 9090, Belgium
Department of Pathology, Bacteriology and Avian Diseases, Ghent University, Faculty of Veterinary Medicine, Salisburylaan 133, Merelbeke 9820, Belgium

Geertrui Rasschaert
Institute for Agricultural and Fisheries Research (ILVO), Technology and Food Science Unit, Food safety research group, Brusselsesteenweg 370, Melle 9090, Belgium

Fabio Granados-Chinchilla
Centro de Investigación en Nutrición Animal (CINA), Universidad de Costa Rica, Ciudad Universitaria Rodrigo Facio, San José 11501-2060, Costa Rica

Sugey Prado Mena
Centro de Investigación en Nutrición Animal (CINA), Universidad de Costa Rica, Ciudad Universitaria Rodrigo Facio, San José 11501-2060, Costa Rica

Lisbeth Mata Arias
Centro de Investigación en Nutrición Animal (CINA), Universidad de Costa Rica, Ciudad Universitaria Rodrigo Facio, San José 11501-2060, Costa Rica
Escuela de Zootecnia, Universidad de Costa Rica, Ciudad Universitaria Rodrigo Facio, San José 11501-2060, Costa Rica